Plant-Inclusive Care Ethics

Examining anthropocentric assumptions and arguments in care ethics literature, this book argues that caring for plants represents a moral obligation and vital relationship between human carers and plant cared-fors.

By exploring the moral value in plants and their capabilities for caring relationships, the book demonstrates that this type of care is already in practice and discusses three case studies of human-plant contexts: bonsai, tomato plants, and giant sequoias. Brelje reasons that plants should be included in care ethics based on the human carers and plant cared-fors ability to meet the core requirements of caring relationship: carers' valuing-attitudes for plants, caring affect towards plants, human-plant interdependency, responsive interactivity between humans and plants, and plant success conditions. This theory is applied to real life caring relationships between human carers and plant cared-fors through interviews with bonsai practitioners, tomato plant farmers, native caretakers, and parks personnel, providing sketches of contemporary human-plant care relationships. Each case offers context-relevant challenges based on plant type and use, aesthetic, and consumption, while highlighting the common ethical care patterns.

Plant-Inclusive Care Ethics will be useful for students, instructors, and practitioners in environmental ethics, plant humanities, critical plant studies, and plant-related artistic work.

Kate Brelje is a researcher at the intersection of environmental ethics and feminist theory. She is Assistant Professor of Philosophy at St. Bonaventure University. Previous publications include "More than Humans: A Case for Inclusion of Non-human Persons in Care Ethics" in *Essays in Philosophy* (2023). She is also the showrunner for the Networking with Plants in the Anthropocene Podcast (soundcloud.com/networking-with-plants).

Routledge Environmental Ethics

Series Editor: Benjamin Hale, *University of Colorado, Boulder, USA*

The Routledge Environmental Ethics series aims to gather novel work on questions that fall at the intersection of the normative and the practical, with an eye toward conceptual issues that bear on environmental policy and environmental science. Recognizing the growing need for input from academic philosophers and political theorists in the broader environmental discourse, but also acknowledging that moral responsibilities for environmental alteration cannot be understood without rooting themselves in the practical and descriptive details, this series aims to unify contributions from within the environmental literature.

Books in this series can cover topics in a range of environmental contexts, including individual responsibility for climate change, conceptual matters affecting climate policy, the moral underpinnings of endangered species protection, complications facing wildlife management, the nature of extinction, the ethics of reintroduction and assisted migration, reparative responsibilities to restore, among many others.

Ecological Ethics and the Philosophy of Simone Weil
Decreation for the Anthropocene
Kathryn Lawson

Virtue Ethics and the Environment
Dominika Dzwonkowska

Positive Duties to Wild Animals
Kyle Johannsen

Environmental Ethics and our Collective Character
David Clowney

Plant-Inclusive Care Ethics
Insights from Bonsai, Tomato Plants, and Giant Sequoia Trees
Kate Brelje

For more information about this series, please visit www.routledge.com/Routledge-Environmental-Ethics/book-series/REE

Plant-Inclusive Care Ethics

Insights from Bonsai, Tomato Plants, and Giant Sequoia Trees

Kate Brelje

Routledge
Taylor & Francis Group
LONDON AND NEW YORK

earthscan
from Routledge

First published 2027
by Routledge
4 Park Square, Milton Park, Abingdon, Oxon OX14 4RN

and by Routledge
605 Third Avenue, New York, NY 10158

Routledge is an imprint of the Taylor & Francis Group, an Informa business

British Library Cataloguing-in-Publication Data
A catalogue record for this book is available from the British Library

ISBN: 978-1-032-99347-8 (hbk)
ISBN: 978-1-032-99348-5 (pbk)
ISBN: 978-1-003-60366-5 (ebk)

DOI: 10.4324/9781003603665

Typeset in Times New Roman
by Taylor & Francis Books

For my family.

Contents

Acknowledgements

This project is a community effort and could not have been completed without the love, attention, and energy of so many.

I extend my gratitude to the team at Routledge, especially series editor, Ben Hale, editors Grace Harrison and Morwenna Scott, and editorial assistant Ananya Singh. I am also grateful for funding from the Temple University Office of Sustainability Temple University Graduate School, and Winterthur Museum, Garden, and Library for financial support of parts of this book project.

To my academic community, I extend my heartfelt gratitude to Miriam Solomon, Eugene Chislenko, Paul Crowe, Katie McShane, Espen Hammer, Laura Levitt, Katie Brennan, Moonyoung Hwang, Mengting Wang, Ziqian Zhang, Yining Wu, Tania Islam, Sediqa Fahimi, Daniel Remer, Grace Cooper, Mary Ward-Bucher, Mia Brett, Alex Gillham, Elizabeth Cantalamessa, Heather McDivitt, Stephen Setman, Russell Woodruff, Kenneth Pearce, Cyril Uy II, and Nancy Klancher for their help navigating this project. Special thanks to my colleagues and friends from Networking with Plants in the Anthropocene, whose conversations and work have inspired and informed me: Paul Moss, Anna Perdibon, Deepta Sateesh, Meera Baindur, Evgenia Emets, Joela Jacobs, Prudence Gibson, Paco Calvo, Natalie Lawrence, Evelyn Brister, Jon Pitt, Thomas Hart, Sydney Kale, Alyssa Paredes, Cate Sandilands, Sophie Chao, Maura Flannery, Yogi Hendlin, Kat Gutierrez, Steven White, Jill Pfluheber, Sury Jha, Madii Kasem, Solvejg Nitzke, Dani Lamorte, Mariko Thomas, Melissa Parks, and Don "Flip" White.

I am also deeply grateful for the conversations that I had with the interviewees for this project. Without their expertise and willingness to share their experience, this project would not have been possible.

Finally, my friends and family have helped me navigate the creation of this book. I am indebted to my friends Gwendolyn Reyes-Illg and Susan Meister, who helped keep me grounded in this project. I am deeply grateful to my parents, Val and Jon, my dog siblings (Mickey, Bandit and DeeDee), and my partner Matt Lamb. I am especially grateful to my more-than-human family members, Aria and Sprite, who are constant companions and have reminded me of what is important in this world.

Introduction

Humans care for and about a lot of things. Frequently, humans care for other humans, like their children, friends, and family. Humans also care about a lot of things, from intangible ideas like freedom to tangible objects like bicycles (Tronto 1993). But can humans care *for* more than human beings? We frequently talk about care in English and use to the term to describe our actions towards other animals (e.g., pet care) and plants (e.g., lawn care). And so, the question of whether we can care for those beings might sound ridiculous. Of course we can, and in fact, we frequently do! But the context of care that I am concerned with in this work refers to ethical care; the care at the heart of care ethics. When the term "care" is invoked, sometimes it relates to actions and attitudes in a non-moral or amoral way. But when it refers to a morally laden relationship, in important ways similar to the care that we talk about when describing our relations with other humans, a deeper analysis of the moral character and stakes of that care is necessary. That is the overarching question of this book: is the caring relationship between human carers and plant cared-fors the stuff of ethics? Or are human-plant relationships outside the realm of care ethics. And if the human-plant caring relationship is ethically laden, what does that mean for our common caring practices with plants?

Care ethics might seem a natural fit to understanding and evaluating our ethical relationships with plants. Not only because of the vernacular use of care in English noted above, but also due to the historical relationships humans have with plants more generally. Narratives about the agricultural revolution in the evolutionary history of human beings emphasize the link between the flourishing of human civilization and the domestication (or at least, intentional cultivation) of plants for consumption and use (Blakemore 2019). Religious traditions (even anthropocentric ones) can describe the proper relationship between humans and environmental entities as one of care (Catholic Church 2015).

And yet, care is fraught. As will be addressed throughout the text, common criticisms of care ethics (e.g., that it is not critical enough of societal gender norms, that it is not robust enough to provide a proper ethical theory, that it can be reduced to more mainstream ethical theories like virtue ethics) arise when we are considering whether it is a theory capable of including more-than-human

DOI: 10.4324/9781003603665-1

entities. More directly, care ethics is already controversial when applied only to relationships between humans—how much more so will it be when stretched to include the more-than-human world?

My interest in this problem emerges, like many feminist projects, from labor and lived experience. As a graduate student, I worked on a municipal seasonal botanical crew as a plant carer. I would drive around in vehicles, large and small, to water planters filled to the brim with annuals. Often, I would have the same route: starting with hanging planters dangling from light fixtures in the middle of parking areas, to planters hanging on the sides of sidewalks, to the large and small circular ones dotting the town square. We would plant beds, set up irrigation, and provide weeding and upkeep. It was quite an investment of time, labor, and money, but the local business owners, residents, and tourists loved the fruit of this labor. Unlike other thankless government jobs, the care and maintenance of these beings was normally a source of joy. The only complaints I semi-regularly received were from transient, unhoused persons who asked why we didn't have more food trees, a concern I admit I also frequently thought about while tending to the often non-medicinal, non-edible, and non-native annuals.

Plant tending was also a constant part of my life. My mother is a gardener, and her father and mother before her, her grandparents before them, and so on. In her lifetime, as a child in a working-class family, their garden was a primary source of sustenance. In my lifetime, as a child of the lower middle class, our garden was a mixture of sustenance and cultural habit. Outside of the spring-to-summer-to-fall outdoor garden, our house was filled with potted plants. Twice a year, we would go through the shift: an exodus of plants to the outdoors for the warm summer, and a fall home-coming of potted tropical plants back into the house. It was a core seasonal flow of the household. In this research, I feel a melding of worlds, a reconnection to my mother's lineage of plant caretakers and engagement with learning and research present in my father's family. In this work, I feel my hands are finally back in the soil, literally and figuratively.

I begin the introduction here, with this personal anecdote and history, for the reader to understand my positionality. Sometimes in feminist research, the author will include how they are connected to their research subjects. As you will see, in each case study, I will include my relationship to the plant chosen to illuminate a given context. This helps complicate the assumption of a god's-eye-objectivity. Especially as a plant caretaker myself, I feel the need to be clear about my orientation and history in approaching this philosophical problem. It is this experience that initially sparked a philosophical interest in plants as beings in-and-of-themselves. Given my situatedness, there may be values and assumptions that I am bringing to this project that are intuitive to me but might be huge steps in logic or unintuitive to readers and interlocutors (Longino 1990). I welcome the critical process of unearthing, describing, and questioning these values (probably due in part to my understanding of both feminist and environmental philosophies as inherently critical projects).

The way that practicing care of plants cultivated in me a greater awareness of plants and concern for their wellbeing made me wonder if others had a similar experience. I thought that humans who work with plants may, on occasion, find sparks of value in that practice. Caring for plants could illuminate not only the moral value and structure of caretaking, but also greater awareness of the value and capabilities of the plants themselves.

As I began to look through the work of care ethicists to understand how plants fit into their moral schemas, the foundational ethicists I examined (primarily Nel Noddings, Eva Feder Kittay, and Virginia Held) either did not have space for non-humans or explicitly stated they were outside the bounds of care ethics. Ethics of care appeared to be concerned with human societies and human relationships, not animals, plants, ecosystems, etc. As I looked more closely at their accounts, though each has a distinct approach to care ethics (i.e., different theoretical structures), each focused on humans or human persons.

Broadening beyond the work of Noddings, Kittay, and Held, many others maintain a focus on humans (see Gilligan 1982 and Ruddick 1989). However, within the explicitly political-ethical intersection of care ethics, Joan Tronto (1993) and those who follow her lineage, like María Puig de la Bellacasa (2017), include ecological entities within their range of care. While Tronto and Puig de la Bellacasa are engaged here, my interest in two things leads me back to Noddings, Kittay, and Held. First, much of the literature surrounding care ethics (both critical and expansive), Gilligan, Noddings, and Held are invoked. As I am attempting to make sense of care ethics applying to a larger realm of beings, engagement with these thinkers who are resistant to it is important, as is evaluating their claims. Related in a generic sense, care ethicists like Noddings are invoked as theories that might have resources for environmental ethics (Hourdequin 2015, 85). There is something attractive about the fit between care ethics and the more-than-human world. I want to interrogate that possibility, whether the ethical theories that appear anthropocentric, can apply more broadly to a wider range of cared-fors beyond humans.

Political ecofeminist projects were what led me to this project, particularly the questions of domination and oppression of nature (Plumwood 1994, Warren 1990). Initially, I was interested in the ecofeminist question of the wrongness of domination over plant life. Whether there is a wrongness-making feature that makes the domination of plants bad may rely on whatever morally relevant capabilities plants have. So, this project of determining why domination of plants is bad developed into an investigation of what plants are and how they fit within normative ethical theories, particularly care ethics. Political accounts, as many ecofeminist projects are, are concerned not only with individual ethical relationships, but larger societal, structural, and institutional forces that contribute to phenomena like oppression. While there are approaches, like Joan Tronto's, that engage the ethics of care with political argument and application, I am focused on a modest project more directly aimed at the ethical framework of care ethics.

Kittay follows the work of Noddings, both arguing for an enriched theoretical account of care ethics and for expanding the realm of possible cared-fors. I think Kittay provides one of the most comprehensive accounts of care ethics as an ethical theory. She engages the question central to my work here: who is a cared-for? Since Kittay is motivated by concerns about inclusion of humans who have been historically gate-kept from consideration as full moral persons, she zeroes in on who can be a cared-for and why. This approach identifies central characteristics that a cared-for must have in order to qualify as a cared-for in a morally relevant way. In considering whether plants can fit into an ethically caring relationship, I too must make sense of what characteristics and capabilities plants have for "elevating" them from beings who are considered worthy of care in a non-moral sense, to beings demanding of care in a moral sense. Along with Kittay, Noddings and Held provide central care ethics theories that engage both the character of care and the structure of normative theory.

To be sure, there are numerous criticisms of care ethics; and many of them will still hold against my project of including plants within care ethics. Early concerns with care ethics surrounded apprehension that care ethics allows no space or place for justice (Hoagland 1990). Many feminist projects are concerned with injustices, like sexism, racism, xenophobia, homophobia, and transphobia. An ethical theory that does not include an account of justice may not provide space for explaining these phenomena and providing solutions for combatting injustices.

Relatedly, there have been charges against care ethics as a feminine ethic, rather than a feminist one. Care could be a part of traditionally feminine gender roles (if there are clear cut definitions of such things). Rather than question the roles that feminized persons are granted (e.g., mothering, caretaking, supporting, laboring for the other), care ethics could be seen as doubling down on or uncritically uplifting these activities that have been used as tools of domination and oppression. They may also create norms that lock people into these roles, if they are determined by one's gender. Extending care ethics to include plant cared-fors does not address this fundamental concern. Though, perhaps it could chip away at the stereotypes of carers as more masculine care activities like farming or forestry could be considered realms of ethical care.

One approach to understanding the intervention of care ethics is to think about who usually does care work and has caring relationships. By making care central, people engaging in care work are made central. Often, but not exclusively, these care workers are women. Stereotypical care work includes child rearing, feeding, bathing, teaching, aiding, and other activities that are associated with women in many cultural contexts. By uplifting this gendered notion of care, it could be an adoration of gender stereotypes or norms that goes unquestioned and uncontested, rather than interrogate how these traits became associated with women and femininity in the first place. And again, not all women are feminine, and not all carers are women. So, an uncritical assumption of gendered characteristics might contribute to misconceptions of care ethics being feminine-dominant (meaning to be feminine and woman is to be superior than masculine or man) or non-critical of social characteristics that may not be innate to any sex or gender, but instead socially constructed. Both of these misconceptions may come from gender essentialism, a view that there are particular characteristics essential to each gender and sex.

These concerns also pertain to ecofeminists, including famously Vandana Shiva. Shiva's ecofeminist projects commonly involve plants, particularly around GMOs and patents, which Shiva claims are dangerous to humans and pernicious to indigenous knowers. Patents by global corporations can be weaponized against traditional practitioners by use of copyright laws prohibiting the use of traditional substances (e.g., neem oil) or by profiting off of this product without proper compensation to the traditional groups that held this knowledge (Shiva 1997). Shiva also asserts what many consider to be gender essentialism: that by their very nature, women are more connected to the earth than men. This amounts to gender essentialism because she is asserting an essential characteristic or quality to all women. This ignores individuals who are women who do not have this quality (e.g., innate and special connection to the earth). Critics of Shiva note her gender essentialism as glorifying the feminine in womanhood does not acknowledge unfeminine women and does not call the gender binary (i.e., women and men, females and males) into question (Gaard and Gruen 1993). So rather than being a feminist ethic, it is a feminine one because it glorifies femininity without providing critical engagement with these social roles and the injustice that they might promote.

These concerns, especially coming from other feminist ethicists, could hold particular weight if a foundational principle of feminist ethics is a concern with sexism and unjustified gender-based exclusion, domination, and oppression. Later responses argue for space within care ethics to make sense of justice and non-essentialist accounts of sex and gender (Spelman 1988, Moody-Adams 1991). And similarly in environmental philosophy, subfields like queer ecology challenge gender and sex binaries, in human societies and the more-than-human (Sandilands and Erickson 2010).

Another general concern about this project, relating to care ethics more generally, is that I am offering an extensionist approach. An extensionist approach is one where a theorist takes an existing theory and tries to extend it to a greater number or kind of entities. So, for example, Tom Regan's work in animal ethics could be understood as an extensionist project. If you interpret his work as taking the concept of Kant's Humanity Formulation, and extending its account of inherent moral value to all subjects-of-a-life (Kant 2002, G 4:429), then he is beginning with an account that originally applies to humans and extending its application beyond the human community. Replacing reason with morally relevant capabilities, he is able to make a case for why more-than-human animals also deserve the type of respect and dignity present in the Humanity Principle in virtue of their inherent moral value (Regan 1983).

This relates to my project as I am working with an ethical theory that is traditionally anthropocentric (i.e., primarily concerned with and only available to humans), and attempting to extend it to plants. This could be problematic on its face for a number of reasons. To begin with, extensionist approaches are usually constructed in and for a different community than they are later applied to. So, the contents and structure of the ethical theory were made with humans in mind. To take that theory and apply it to other animals or even plants may seem like a non sequitur. I hope by the end of this work you'll concede that it is not. But it is a reasonable initial concern about any extensionist project. Another concern is about the community the theory is being extended to. If care ethics is a theory comprised by humans for humans, then am I doing a disservice to plant life by trying to force them into my own all-too-human mold? There is a concern that by emphasizing similarities or throughlines, that one can lose sight of the unique, more-than-human beings that are plants. Plants are not humans and they do not need to be humans to matter. But by extending a human-centered ethic to include plants, it might appear as if that is exactly what the theorist is rebutting; that plants must be human-like in order to matter. That is why we start with ethical theories developed in the human world rather than ones based on relationships and interactions with the more-than-human ones.

Key to environmental ethics more broadly is an engagement with science. As an applied ethic, environmental ethics frequently engages with the real world; that requires engagement with the sciences. Within the context of a plant-inclusive care ethic, plant science is important, because determining how or why plants may be morally valuable is connected to their actual capabilities and capacities. Consequently, this project does engage some plant scientists (e.g., Anthony Trewavas, Stefano Mancuso, Lincoln Taiz, Daniel Chamovitz and Francis Hallé) to better ascertain what those capabilities are. In contemporary plant sciences (as with many other areas of scientific research), there is peer disagreement about what these capabilities are and how to describe and study them.

As a philosopher working in an interdisciplinary space on an applied ethical project, I must be clear about my expertise and its limits. I am not a plant scientist or a social scientist (this will be particularly important and explained in Chapter 4) but my work requires engagement with scientific claims about plant capabilities and ethnographic studies into human perspectives on their relationships with plants. As you will see, I understand my position to be one of flagging difference and disagreement, and acknowledging what that might mean for a view. Rather than argue for one particular option of including plants within care ethics, I offer a range (see Chapter 2). In doing this, I am advancing a philosophical argument of how plants could count as cared-fors within an ethical framework, and acknowledging what support could be offered for each view. Ultimately, I pursue a value pragmatist view of which source of moral value an individual chooses to admit. But for each, I attempt to provide reasons offered in the literature for accepting or rejecting each view. In this way, I hope to avoid masquerading as the final arbiter of ongoing scientific disagreements over plant capabilities and their descriptions.

This issue opens a reoccurring concern that will appear throughout this work: anthropomorphism. Anthropomorphism is roughly the ascription of human characteristics, traits, or values to more-than-human (or non-human) beings and things. As a quick example using a non-human thing, when I drop my cell phone, I wince. For me, the wince is not a fear of how much it will cost me to fix it, or the initial concern about inconvenience if the phone is seriously broken. My wincing is a type of projection of feeling (sentience) to my phone, despite the inability of my phone to experience pleasure or pain. It is involuntary and a little silly (what can I say? I am occasionally an overly sentimental person). One might argue an unreflective attribution of a human quality (sentience) to the non-human (a machine) is an instance of anthropomorphism. And in the case of my cell phone, such an attribution is false; my cell phone does not experience pain upon the cracking of its screen. Anthropomorphism gets trickier when we ascribe things that humans feel to other animals, and in some cases, other living beings or natural entities. Once upon a time, attributing certain types of sentience to other animals (e.g., interpreting my dog's sharp bark as an expression of frustration over his inability to access the outdoors immediately and independently) would be laughable. Hopefully this is no longer the case. But ascriptions of complex relationships that humans experience (like friendships) to other animals (especially ones that we use for food, like cows) can still be read as a type of anthropomorphism, despite evidence that some animals do appear to create and maintain preferred social relationships (Gutmann, Špinka, and Winckler 2015). If someone is comfortable with cows having best friends, they may not grant that plants can be. Plants tend to be so different from us, both as humans and as animals, that attributions of concepts or qualities that occur with humans and other animals to plants is a nonstarter. Yet, some humans do this! They describe plants as happy or sad, flourishing or suffering. A good amount of work being done in the plant humanities and critical plant studies right now is around anthropomorphism, and when ascribing human-like capabilities to plants can be justified, as well as when it can't. It muddies the water on whether anthropomorphism is always bad or always good. Surely the ascription of formerly-reserved-only-to-humans-capabilities or relationships to some other animals can be beneficial, true, and justified. Perhaps this can also hold, under scrutiny, to some plants and natural entities.

This project is conservative in that it does not address fascinating questions of plants as carers, or caring relationships comprised only of more-than-human beings. My primary philosophical concern is making space for plants as cared-fors within an ethic of care. In this book, I contend that plants can and should be considered cared-fors. In future work, I hope to address other questions related to this project more in-depth, such as plants as carers, plants as social beings, and plants as world-makers. But for now, I address the issue of what someone who wanted to admit plant life into an ethics of care would need to do to produce a coherent, inclusive care ethic.

A central issue in this extension of care ethics to plant cared-fors is the question of plant capabilities. There is a wide variety of beliefs about the identity of plants and their abilities across the fields (as well as intra-disciplinarily) of critical plant studies, plant humanities, the plant sciences, and religious traditions. Take the example of yagé, *Banisteriopsis caapi*. Yagé is a vine native to Amazonian region. Most often, it is utilized by humans for its psychotropic qualities. It is most well-known in an anglophone context as ayahuasca, even though this can refer to a concoction or process of consuming a number of different plants (Pfluheber and White 2025). Experts in critical plant studies examining this facet of the yagé plant may emphasize the claims of plant capabilities in producing particular visions and describe the plant as being a communicative agent. A plant humanist might focus on the art, writing, and oral stories associated with yagé and describe the plant as being a culturally important or even active. At the same time, the plant physiologist/botanist/biologist would explain the same activity in terms of chemical interaction within the human body. Finally, the religious practitioners who utilize yagé in a ritualized setting may explain the plant's capabilities as connected to a spiritual plane. That different fields have different explanations and definitions, according to their goals and methodologies, is nothing new or surprising. It becomes an issue for interdisciplinary projects like this one when holding multiple interpretations of a phenomenon. And so, in deciphering the testimony of plant carers about the capabilities of their plants to make sense of the contours of their caring relationships, there is an imperative to both understand areas of disagreement within the sciences and cultural beliefs and also attempt to make sense of these differences in a way that could clarify points of non-contradiction. There are so many examples of this work being done in the plant humanities and critical plant studies right now (Gagliano 2018, Gutierrez 2025, Pitt 2025, Sprinkel 2025). They provide a curious and engaging juxtaposition of philosophical, spiritual, and cultural meaning next to contemporary science.

Beyond this difference in capabilities ascribed and allotted by different disciplines, plant life is incredibly varied (Hallé 2002). In terms of life span, some plants are annuals who go through their life cycle in a single year. Others are biennials, which live for two years. Perennials live for longer and "come back" to life after hibernating in seasonal systems. The trees discussed in this book, both potted and wild, can live for centuries, and in some cases millennia. Plants can be tiny (like tiny baby succulents) or ginormous (like the baobab tree). The diversity of plant life is profound and difficult to fully appreciate in this short introduction. I attempt to provide some acknowledgement of the diversity by including case studies that range from bonsai trees to tomato plants to giant sequoia plants. Human categorization and scientific understanding of plants has also changed greatly over time. For example, humans have changed their consideration of algae, moving them from the Plant kingdom to Protista. Within the tomato chapter, we'll see that plants can transform in popular culture and science from poisonous to help affirming medicine in a short period of time. Environments can also bear on both individual plants and whole communities of plants in where, when, and how genes are expressed. So an individual may develop in different ways, depending on factors that occur in their environment, like temperature, access to water, access to sunlight, exposure to wind, animal interactions, etc.

Applied ethics holds a tenuous space. Normative ethical theories, by their very nature, aim to apply broadly to a diverse set of phenomena. That is how the definitions of good and bad, right and wrong which they provide are able to guide one's actions, amidst changing contexts. But applied ethics assesses the strength of those theories, evaluating whether they hold when connecting with actual situations. In the inevitable grey area that emerges, the applied ethicist carefully navigates the murkiness, examining a myriad of possibilities. Does the challenge of an on-the-ground application require revisions to the normative theory? Does the theory fall so short in adequately evaluating the situation that it must be abandoned for an alternative? This possibility looms in the background of this book. However, I aim to convince readers that care ethics can successfully apply even in these difficult contexts.

Essential in this work is also the actual practice of plant care taking. If it is theoretically sound to include plants as cared-fors, following the example so beautifully offered in Marion Hourdequin's *Environmental Ethics: From Theory to Practice* (2015), practice is also important to address. After arguing for a theoretical project that includes plants as cared-fors, I apply that theory in case studies representing three distinct contexts of caring relationships between humans and plants. These case studies take place in vastly different contexts which test the theory in different ways, namely, aesthetic, consumptive, and wild.

The theoretical portion of this project concerns what I argue are shortcomings of contemporary care ethics that could be remedied by expanding the realm of moral considerability. In Chapter 1, I introduce care ethics and its relationship to plants. Tracing the genealogy of care ethics from Carol Gilligan to Nel Noddings, Eva Feder Kittay, Virginia Held, and Joan Tronto.

I detail the identity-relevant requirements for cared-fors in the care ethics of human-centered accounts of care (Noddings, Kittay, and Held). Examining each account's use of anthropocentrism within the theoretical framework and their arguments for restricting the domain of care ethics to humans, I identify what benefit this human-centered approach serves. Anthropocentrism shows up in two primary ways within care ethics: first, by setting apart relationships between humans from relationships humans have with other beings; and second, by granting intrinsic moral value only to humans, or significantly enough human-like beings. After making sense of the work anthropocentrism is doing in each theoretical structure, I argue against this limitation of the moral community to only humans. These ethical frameworks could remain coherent, even with the inclusion of non-human cared-fors (particularly for my focus, plant cared-fors).

Chapter 2 takes up the question of how an ethicist could include plant cared-fors in these ethical care frameworks. In order to elevate plants from the realm of descriptive care to prescriptive care (i.e., how things in the world are to how they should be), there are two important elements that should be accounted for in plant cared-fors: (1) intrinsic moral value and (2) the capabilities that are relevant for the caring relationship. Having inherent moral value means that the plant cared-for is morally valuable in and of itself, beyond mere instrumental value, and ensures that the plant cared-for belongs in the realm of moral considerability. In the work of Noddings, Kittay, and Held, both the cared-for and carer have inherent moral value. Their identity as humans, or human persons, entitles them with such moral value. Because care ethics is a relational ethic, it also seems important that those involved in the relationship have capabilities relevant to and required in a caring relationship. Considering these requirements, I offer five possible options for plants that would qualify them as cared-fors:

a assert that plants are persons in the traditional definition of moral personhood;
b establish that cared-fors are subjects-of-a-life, a concept from Tom Regan develops to expand the realm of moral patients;
c insist that plants are valuable as living beings;
d explain plants' inclusion with care ethics through their ability to flourish; or
e include plants in virtue of their being kin.

Options (c), (d), and (e) offer an account of a non-person, non-subject-of-a-life cared-for within care ethics. Utilizing a literature review of plant ethicists and environmental philosophers, I examine these options in search of viable sources of intrinsic moral value for including plant life within an ethics of care given contemporary cultural and scientific views of plants. If one grants that plants are able to communicate, for example, that could be very relevant to their ability to participate in a caring relationship. Additionally, if plants are able to sense pleasure or pain, this ability would likely have meaningful bearing on how life goes for them. Sometimes these capabilities are connected not only to determining who or what has intrinsic moral value, but also whether they have the capabilities necessary to be in a caring relationship.

In Chapter 3, I address how plant cared-fors can adhere to key requirements of cared-fors in care ethics. I identify three particularly important components that an ethically caring relationship should account for:

1 an attitude that ascribes inherent moral value to the cared-for,
2 a relationship between the carer and the cared-for that includes caring affect, dependency, and responsive interactivity, and
3 consequences from the actions of the carer that could be beneficial for the cared-for (i.e., success conditions).

I argue for these components of a plant-inclusive care based on the work done in the first chapter surveying central features of ethical care theories. Beyond developing this account, I also consider objections or concerns to the inclusion of plant non-persons within a care ethic. If a plant qualifies as a cared-for, what does that mean for the sheer number of entities we, as human moral agents, are expected to care for? What about the apparent paradox between ethical care and killing and consumption? These and other criticisms and upshots are explored in this chapter.

The next chapters examine distinct contexts of care between human carers and plant cared-fors. I begin in Chapter 4 with an introduction to the normative case studies that comprise Chapters 5, 6, and 7. In the case studies chapters, I use the content of interviews with practitioners to inform whether the account I propose in Chapter 3 can cohere with current practices of plant care in aesthetic, consumption, and wild contexts. The case studies might feel unfamiliar to some philosophical audiences and incomplete to some social science audiences. They are highly interdisciplinary, in that they employ interviews with practitioners in what might look like ethnographical studies. However, as I explain in Chapter 4, they are normative case studies, working as a test case for the theory developed in the first half. Not unlike other applied philosophy work, one goal is to bring the theory into conversation with the world. Some might argue that in order to prescribe how we should care for plants, it is useful to understand how we currently interact with plants. Ultimately, I am curious not only whether an ethical theory is logically possible, but just as important in an applied sense, whether it is practically tenable. This project occurs at a time when there are varied paradigms underlying beliefs about what plants are and what they can do. As the interviews show, within the fields discussed in three case studies (bonsai, tomato plants, and giant sequoias), individual practitioners have a wide array of beliefs about the value of plants, the nature of their relationships with plants, and what it means to benefit the plant through their actions. Part of applied philosophy is taking one's theory out of the ether and applying it on the ground, in part to determine whether it makes sense. I contend that these case studies demonstrate that in varied contexts, an ethic of care that includes plant cared-fors is both coherent and practical.

I decided on these case studies for a few reasons. First, I wanted to highlight a variety of contexts for human-plant care relationships. The bonsai case study provides a largely aesthetic context also full of claims of familial and generational relationships between humans and plants. Tomato plants are grown primarily for consumptive aims, as annuals in climate regions where they are unable to grow natively as perennials. Giant sequoia trees live in apparently wild contexts (though, as the case study will bear out, they also resist the bifurcation between wild and non-wild). Secondly, I wanted case studies that were familiar enough to a wide range of readers. While bonsai might not be a common practice, many people are aware of bonsai and bonsai plants. Tomatoes are a commonly consumed fruit in many cultures around the globe. And giant sequoia trees, while some might view as an oddity, are familiar to many in part because of their large size. The third consideration is the presence of living practitioners in relation to these plants within my socio-linguistic context. The English-speaking bonsai community is large and accessible. Tomatoes are a popular fruit in contemporary cuisines, home gardens, and industrial farms. Giant sequoia trees are still present in the western United States as wild-identified plants that many people in the English-speaking world are aware of, even if they have not interacted with one in person. So, each of these plants is at least vaguely accessible to the contemporary imaginary. Thus, many readers will be vaguely familiar with the plants at the center of these case studies. Less academically importantly, I wanted to focus on plants that hold a place in my own personal context. Each plant, as discussed in the chapters, has some connection to my life. I emphasize the U.S. context of the practices because that is where I am currently writing from. If ethicists are going to apply a theory to the world, I think there is some value in applying it close to home.

The first of these chapters addresses contemporary American bonsai practices. I approach Chapter 5 as a normative case study concerned with care ethics in contexts rich with aesthetic claims and meaning, alongside generational relationships between individual plants and multiple human carers. Bonsai is a controversial practice. Some humans respond aesthetically to the miniature plants with pleasure, others with disgust. Highlighting the moral components of these aesthetic and generational contexts, I offer a brief definition and history of bonsai before engaging with conversations had with contemporary practitioners about their relationships with bonsai plants. These practitioners fall into roughly three groups: hobbyists, professional practitioners (e.g., artists, teachers, craftspeople), and curators. There are interesting ethical questions not only around the application of the theory to the care of bonsai plants, but also in whether bonsai as a practice can ever have good outcomes, or success conditions, for the plants. There is a concern that the process of bonsai itself is not compatible with care. Chapter 5 addresses these concerns in concert with interviews covering a wide range of practitioners.

Chapter 6 addresses what I suspect is one of the most difficult contexts for an ethic of care to engage: one in which the cared-for is raised for consumption. Domestication and reciprocity factor heavily in my analysis of this case study on tomato plants. One area of perennial concern and attention for the consumption context is valuing-attitudes. Practitioners can run afoul of the care ethic if they fail to hold appropriate valuing attitudes towards their tomato plant cared-fors.

The inquiry highlights three sub-groups within the tomato growing communities: farmers, gardeners, and seed-savers. This case study culminates in the pressing question: can we ever successfully and properly care for beings which we tend with the goal of consuming? Or does a context of consumption undermine any possibility of being in a good ethically caring relationship with something?

The final case study is on giant sequoia trees. Chapter 7 introduces this context as a caring for wild plants account. Giant sequoias are an interesting type of wild plant, especially given their symbolic importance within U.S.

culture and history. For this case study, I interview caretakers who work with giant sequoias through governmental organizations and nonprofits, as well as local Native and non-Native individuals who have relationships with the trees. This case study, while focusing on wild contexts, puts pressure on that distinction itself. Within care ethics, the ideal of care (e.g., within family systems and associated with a particular feminine identity) contains an air of domesticity. Using wild and domestic in the common dichotomous fashion, non-human, "natural" entities are often grouped into one or the other of these categories. However, growing evidence suggests that the impulse to view things as either wild (i.e., distinct from and devoid of intervention from humans) or domestic (i.e., managed by and intervened upon by humans) might be both historically mistaken and philosophically problematic (Plew and Yu 2025). This case study brings these issues to light within the context of a care ethic.

Throughout this exploration of care ethics, plants and their carers take a central role. I think it is important to engage directly with carers and their work because environmental ethics, ecofeminist or otherwise, is an applied ethical tradition. Especially in light of the care ethic perspective that this analysis engages, relationships are essential, foundational, and informative. To better understand the contours of care, following from the work of Kittay and others, I want to engage with the lived experiences and testimonies of plant carers. Engagement in this context means navigating tensions of deference to expertise and lived experience while responding to the requirements of a normative framework of care ethics.

Grounding this research in relationships, I aim to work within the lineage that I find so compelling about both feminist philosophy and environmental ethics. I consider both subfields to be pursuing the call of philosophy in their very essence: critical reflection and engagement with mainstream assumptions and canon. Feminist philosophy emerges in response to a field filled with canons that do not reflect the concerns and identities of many philosophers. Questions of gender, sex, and oppression arise when examining this exclusion. Similar concerns about the exclusion of the more-than-human world emerge from the examination of these canons by environmental ethicists. This critical edge is essential for philosophy, and can be prompted by such seemingly innocuous questions as "Where are the women?" (Enloe 1987) and "What about nature?" (Rolston 1994). So, here the project of critical engagement is continued as we ask, "What about plants?"

References

Blakemore, Erin. "The Neolithic Revolution—Facts and Information." *National Geographic* (2019). www.nationalgeographic.com/culture/article/neolithic-agricultural-revolution.

Catholic Church. *Laudato Si': On Care for Our Common Home*. Our Sunday Visitor, 2015.

Enloe, Cynthia. *Bananas, Beaches and Bases: Making Feminist Sense of International Politics*, 2nd edn. University of California Press, 2014.

Gaard, Greta and Lori Gruen. "Ecofeminism: Toward Global Justice and Planetary Health." *Society and Nature*, 2 (1993), 1–35.

Gagliano, Monica. *Thus Spoke the Plant: A Remarkable Journey of Groundbreaking Scientific Discoveries and Personal Encounters with Plants.* North Atlantic Books, 2018.

Gilligan, Carol. *In a Different Voice: Psychological Theory and Women's Development.* Harvard University Press, 1982.

Gutierrez, Kathleen Cruz. *Unmaking Botany: Science and Vernacular Knowledge in the Colonial Philippines.* Duke University Press, 2025.

Gutmann, Anke Kristina, Marek Špinka, and Christoph Winckler. "Long-term Familiarity Creates Preferred Social Partners in Dairy Cows." *Applied Animal Behaviour Science*, 169 (2015), 1–8.

Hallé, Francis. *In Praise of Plants.* Translated by David Lee. 2002.

Held, Virginia. *The Ethics of Care: Personal, Political, and Global.* Oxford University Press, 2006.

Hoagland, Sarah., "Review: Some Concerns about Nel Noddings' 'Caring'." *Hypatia*, 5(1) (1990), 109–114.

Hourdequin, Marion. *Environmental Ethics: From Theory to Practice.* Bloomsbury Publishing, 2015.

Kant, Immanuel. *Groundwork for the Metaphysics of Morals.* Translated by Allen W. Wood. Yale University Press, 2002.

Kittay, Eva Feder. *Learning from my Daughter: The Value and Care of Disabled Minds.* Oxford University Press, 2019.

Longino, Helen E. *Science as Social Knowledge: Values and Objectivity in Scientific Inquiry.* Princeton University Press, 1990.

Moody-Adams, Michele M. "Gender and the Complexity of Moral Voices." In Claudia Card (ed.), *Feminist Ethics* (pp. 195–212) University of Kansas Press, 1991.

Noddings, Nel. *Caring: A Relational Approach to Ethics and Moral Education*, 2nd edn. University of California Press, 2013.

Pfluheber, Jill and Steven F. White. *Microcosms: Sacred Plants of the Americas.* Papadakis Publisher, 2025.

Pitt, Jon L. *Botanical Imagination: Rethinking Plants in Modern Japan.* Cornell University Press, 2025.

Plew, Mark G. and Pei-Lin Yu. "Perspectives on Early Amazonian Agriculture from Guyana and Venezuela." *Quarternary*, 8(3) (2025).

Plumwood, Val. *Feminism and the Mastery of Nature.* Routledge, 1994.

Puig de la Bellacasa, María. *Matters of Care: Speculative Ethics in More than Human Worlds.* University of Minnesota, 2017.

Regan, Tom. *The Case for Animal Rights.* University of California Press, 1983.

Rolston, HolmesIII. *Conserving Natural Value.* Columbia University Press, 1994.

Ruddick, Sara. *Maternal Thinking: Towards a Politics of Peace.* The Women's Press, 1989.

Sandilands, Catriona and Bruce Erickson (eds). *Queer Ecologies: Sex, Nature, Politics, Desire.* Indiana University Press, 2010.

Shiva, Vandana. *Biopiracy: The plunder of nature and knowledge.* South End Press, 1997.

Spelman, Elizabeth V. *Inessential Woman: Problems of Exclusion in Feminist Thought.* Beacon Press, 1988.

Sprinkel, Olivia. *To Hear the Trees Speak: A Memoir of Nature, Resilience, and Deep Listening*. Bedford Square Publishers, 2025.

Tronto, Joan. *Moral Boundaries: A Political Argument for an Ethic of Care.* Routledge, 1993.

Warren, Karen J. "The Power and Promise of Ecological Feminism." *Environmental Ethics*, 12(2) (1990).

1 What About the Plants?

Plants in Human-Centered Ethics of Care

Introduction

The centrality of care in plant-tending is reflected in common English phrases like "garden care," "lawn care" and "plant care." Closely related words to care are also used frequently when describing the work humans do with plants: cultivate, tend, nurture. Yet, prominent philosophers Nel Noddings, Eva Feder Kittay, and Virginia Held exclude plants from their care ethic frameworks. One might wonder if the vernacular care language is merely a non-moral description of human relationships with plants. The activities that humans engage with in their gardens and on their terraces, in their living rooms and on their farms merely refer to caring actions that are devoid of the moral quality found in care ethics. Instead, it is simply a useful description of care actions that does not connect with a larger normative ethical theory. This may be the case. However, given advances in environmental ethics, ecofeminism, and the plant humanities, it is worth taking a closer look at these practices in relation to care ethics to ascertain whether plants could (or should) be considered proper cared-fors within a moral care schema.

Care ethics centers the relationship between a carer and a cared-for. One locus of anthropocentrism within these theories is apparent in the assertion that humans can be cared-fors, but other animals, plants, and other non-human entities cannot. Anthropocentrism can also be described as something that is centered on or around humans, as a measuring stick of sorts. Both of these types of anthropocentrism can be connected to what Marcello Di Paola terms "moral anthropocentrism" (Di Paola 2018). In one way or another, these facets of anthropocentrism assert some "superiority" of humans in morality, over and against all other beings. Some might argue that the resistance to accept more-than-human beings as carers is also an extension of anthropocentrism. While I think the rejection of more-than-human beings as carers is open to reasonable argument (though I will not offer such an argument here), I do not think that it is necessarily a case of anthropocentrism in the sense that I am using here. In the strictest sense, anthropocentrism in the environmental context refers to humans as the sole members of the moral community, in terms of possessing intrinsic moral value. While the ascriptions

DOI: 10.4324/9781003603665-2

of qualities that a carer must have as a moral agent (i.e., an individual who is capable of understanding morality and being held responsible for their actions, Regan 1983, 15–52) are generally attributed to some humans and denied to more-than-humans, this in itself does not deny that more-than-humans could have moral value. Rather, it denies that they have the capabilities to act as agents in a moral sense. Some might consider this to be inaccurate or unjust, since some other animals appear to adhere to certain rules of engagement in social groups and understand a type of proto-morality (Bekoff 2007, De Waal 2016). But the denial of the capabilities that comprise moral agency does not automatically mean that such a being could not have intrinsic moral value.

Given my goal to consider plants as cared-fors within care ethics, my primary use of anthropocentrism refers both to the view that all humans and only humans have intrinsic moral value, and that the focus of ethics should be on humans (and for care ethics, relationships between humans). So, the theories of Noddings, Kittay, and Held can be considered anthropocentrist because only humans have inherent moral value or are treated as valuable in and of themselves in the normative theory, and because the primary focus of the theories is on humans only. It is important to note that some contemporary environmental ethics have moved away from discussions of moral value. Popular in the early days of environmental ethics (Goodpaster 1978, Taylor 1981, Rolston 1988, Callicott 1999), discussions of who or what matters can be unfashionable—in some cases for good reason. For example, regardless of whether one espouses anthropocentrism (i.e., all humans have intrinsic moral value), sentiocentrism (i.e., all sentient creatures of intrinsic moral value), biocentrism (i.e., all individual living things have intrinsic moral value), or ecocentrism (i.e., all ecosystems or components of ecosystems have intrinsic moral value), the prescribed environmental action may be the same.

There are pragmatist arguments that urge us to look beyond squabbles over who or what matters to what needs to be done (Katz and Light 1996). Climate crises that no longer appear looming, but current, require similar actions regardless of whether you think only humans have intrinsic moral value or who ecosystems do (Fesmire 2020). And all of the positions from anthropocentrism to ecocentrism morally require us to do something more than what we're doing now. They all demand a change of course, or an intervention, that improves human activity on Earth. Even were we only to care about and value human lives, morality demands drastic changes in business as usual (Kopnina et al. 2021). The argument goes that time spent arguing that dandelions have one type of value rather than another are inconsequential in this moment, as all of the dandelions (and dogs, and humans, and bumblebees) are threatened by inaction. If the house is on fire, why pause for a debate over what in the house is most valuable and why? Run to grab what and who you can and put the fire out.

However, these concerns over who or what has moral value persist (McShane 2007, Batavia and Nelson 2017). And the type of moral value something (or someone) has can carry radical implications for what should or should not be done to or with that thing (or person) under different moral theories. A large concern of mine within this project is the moral community. Relationships with others in the moral community can look, feel, and function very differently than those outside the community. And in care ethics, relationships are at the center of morality. Understanding who or what we can be in relationship with under the framework of our moral theories is essential not only to understand the realm of moral considerability, but also to guide our actions in the world.

While this chapter establishes evidence of this anthropocentrism, it also seeks to provide a better understanding of the diversity of care ethics and the role that anthropocentrism plays in these theories. By identifying how core theorists promote limiting the role of cared-fors to humans, I am able to determine whether anthropocentrism can be effectively challenged and replaced or if it totally undermines the promise of extension. My argument begins by tracing a brief genealogy of care ethics, introducing both central theories and their reception. I also provide crucial interventions from animal ethicists and ecofeminists. I then introduce personhood, a concept that is useful to consider for limiting the community of ethical care to humans in the framework of three mainstream care accounts (Noddings, Kittay, and Held). Despite having different theoretical structures, in the strictest normative sense, each requires caring relationships to occur between human carers and human cared-fors. The carer is a moral agent and the cared-for is a moral patient, an entity that can be affected by the actions of moral agent (Regan 1983, 15–52). While the question of other animal and plant carers is an interesting challenge, I am most concerned with the identity of the cared-for. It seems much less controversial (though still controversial enough) to include plants as cared-fors, rather than both carers and cared-fors. To demonstrate the underlying assumptions within Noddings, Kittay, and Held, I offer a detailed examination of the identity of the cared-for in each theory. In some of today's dominant cultural paradigms plants are not considered to be persons, and in many cases, they are not even ascribed inherent moral value (i.e., they are only valuable in so far as they support human life). Even if care ethicists extend personhood to humans on the margins (e.g., Kittay), restricting ethics of care to humans reveals a position of anthropocentrism that must be challenged to admit more-than-human cared-fors.

An Introduction to Care Ethics: A Brief Genealogy of Care Ethics and its Reception

Care ethics is often traced to a work in psychology by Carol Gilligan about moral reflection and judgment (Gilligan 1982). *In a Different Voice* provides insight into moral judgments based on gender through analyzing interviews. She argues that descriptively, different genders espouse different moral values and moral decision-making methodologies (e.g., prioritizing relationships for women and principles for men). Initial responses to Gilligan's findings were mixed. Some theorists identified this as a monumental shift in the way we think about moral judgment and saw a space for the development of a morality with feminine (or feminist) values at the center. Gilligan's work led to developments in philosophy on the construction of a distinct ethical approach: care.

Among the first theorists who took up the project of making a more women-centric ethic were Nel Noddings, Joan Tronto, and Virginia Held. Noddings expanded the ethics of care into an applied ethic within philosophy, especially philosophy of education (Noddings 2013). Joan Tronto developed her account of care ethics in conversation with political theory, highlighting structural factors and institutions that could be reformed with an ethics of care (Tronto 1993). Virginia Held developed care ethics as a normative ethical theory distinct from others (e.g., virtue ethics) and centered the value of care (Held 2005).

In waves of responses to these early theories, there were both further developments and critiques. Eva Feder Kittay, a major theorist engaged in this project, expands on the theory developed by Noddings, making it a more fully fledged normative ethical theory—though differing from Held, one compatible with a virtue ethics framework (Kittay 2019). Sara Ruddick expands a care ethic specifically into the work of mothering, examining power and peace (Ruddick 1989).

Historical critiques commonly emphasize the perceived conflict between care and justice (Hoagland 1990). Sarah Hoagland (1990) argues that an emphasis on care undermines valuable moral components like justice, which make sense of the wrongness of oppression present in sexism and gender bias. Claudia Card (1990) critiques the lack of challenge of social hierarchies and institutions that promote and maintain gender-based discrimination implicit in care. Rather than challenge feminized accounts of women as carers, care ethics could be seen as glorifying an essentialist compatibilism, where women are praised for their essential and necessary feminine characteristics and men are praised for their masculine ones. This does not ring of gender justice but seems to be a doubling down on pernicious and oppressive difference.

There are also care ethicists who explicitly engage the more-than-human world. Lori Gruen (2015) emphasizes empathy as a point of connection and expansion of a care ethic that could include other animals. Josephine Donovan and Carol J. Adams (2007) examine the possibilities and role of care ethics in the realm of animal ethics. Deane Curtain (1991) puts forward an ecological ethic of care. In *Matters of Care*, Puig de la Bellacasa expands Tronto's work to whole ecosystems (Puig de la Bellacasa 2017). Puig de la Bellacasa describes her work in Matters of Care to be "speculative ethics" in the posthuman context, "expanding on care by moving the investigation of its meaning into a mostly uncharted terrain: the meanings of care for knowing and thinking with more than human worlds in technoscience and naturecultures" (Puig de la Bellacasa 2017, 12), following in the lineage of Donna Haraway.

My work here is more embedded in human worlds. I appreciate and am fascinated by the work of feminist scholars who engage ethics alongside political philosophy in conversation with the sciences and posthumanism. Despite my love and passion for the more-than-human world, I'm not quite in the space of the posthumanists. I want to engage more deeply with conventional views to see if bridges can be built with existing frameworks before delving too creatively into alternative approaches.

The Conceptual Core of Care Ethics

Care Ethics is a rich subfield of normative ethics that centers on caring relationships in our lives. When determining what is good and bad, or right and wrong, ethics of care emphasizes the relationships that make our human lives possible. As deeply social animals, we are dependent on others for our very physical existence, as well as social and moral development. Human beings require care. So, rather than grounding ethical theory on the ideal of an individual rational adult moral agent, capable of moral understanding and the sovereign freedom to act accordingly, care ethics focuses on relationships: parent-child, friendships, partnerships, and a wide range of others. Care ethics takes seriously these relationships, which are rich with affect, dependency, and reciprocity.

The structure of each care ethic differs. Noddings, a foundational voice in the ethics of care literature, construes care as a type of relationship, while Kittay proposes that care ethics has three primary facets: labor, attitude, and virtue. Held holds that care ethics is its own type of normative theory, divergent from the structure of virtue ethics, centering care as a value and practice.

Personhood

In some contemporary environmental contexts, anthropocentrism can carry with it negative connotations, as if it is merely an unjustified rejection of the value of everything that is non-human. Under those environmental accounts, limiting care ethics to human cared-fors is ultimately a mistake because the reasons used by these theorists to differentiate moral value of humans against other beings (e.g., species membership, offensiveness of moral status granted to non-human cared-fors, synonymity of "human" and "person," and the human-centered capabilities required of cared-fors) do not hold up against scrutiny. In the scope of my larger project, the restriction of care to human persons needs to be refuted if there is any possibility of utilizing the conceptual tools of care ethics to advance an account that includes plants. We need to understand what restricting the realm of ethical care is doing within the theory to determine if and how plant cared-fors can belong.

Personhood is a concept keenly tied to humanity. As Daniel Dennett notes, "We normally … treat humanity as the deciding mark of personhood, no doubt because the terms are locally coextensive or almost coextensive"

(Dennett 2017, 287). While it is common to use the terms "human" and "person" interchangeably in everyday English, personhood as a philosophical concept differs from humanity.

The definition of personhood shifts slightly depending on the literature within which it appears. For example, within the Indian Buddhist philosophy of mind discourse, personhood acts as a form of individuation against the concept of self, another philosophically fraught concept tied more to consciousness and ownership of experience (Siderits 2019). In that context, personhood is associated with human bodies (which are in the Buddhist context, ultimately a useful fiction) that individuates one human organism from another over time (Siderits, Thompson, and Zahavi 2011). The connection—and sometimes confusion—between the terms "person" and "self" also demonstrates the relationship between humanity and personhood. African and Confucian philosophical accounts of personhood emphasize relational identity. For example, Motsamai Molefe argues that the Ubuntu "aphorism 'A person is a person through other persons'" (Molefe 2020, 10) means that persons are relationally dependent on and determined by others. In the Confucian context, the identity of persons is comprised of a bundle of relationships with other humans (Hourdequin and Wong 2005, 21).

There are also narrative accounts of personhood. Marya Schechtman's (2014) "person" actively engages in crafting one's narrative identity. This requires persons to have complex cognitive capabilities that can craft a story about self over time. Within Kantian ethics, humans and persons again seem to come together, as seen in Kant's humanity formula, the moral principle protecting the dignity of persons (Kant 2002, G 4:429). This tradition emphasizes the cognitive capacity of reason as the primary indicator of an entity's inherent moral value or full moral status (Gruen 2011, 62–3). Dennett offers an account of personhood with the aim of connecting the metaphysical conception of personhood (i.e., the definitions that are concerned with identity, consciousness, selfhood) and the moral conception of personhood (i.e., moral status and moral considerability) (Dennett 2017, 289). These accounts well establish the tradition of limiting the domain of persons to some human beings.

While these metaphysical concerns clearly bear on the moral ones, moral personhood is most concerned with who or what has moral status and why. This opens the possibility of personhood beyond humans. For example, Mary Anne Warren, arguing within the context of the ethical permissibility of abortion, begins with the decoupling of the terms "human" and "person." She claims that they refer to importantly different things; "human" refers to the individual biological organism that falls within the species *Homo sapiens*, and personhood is a moral signifier (Warren 1973, 53–4). "Human" is a descriptive term setting apart members of a certain type of animal, and "personhood" refers to an entity that has a certain moral status, or moral considerability. For Warren, personhood is determined by five characteristics (consciousness, reason, self-motivated activity, the capacity to communicate, and self-awareness) and admits some other animals, while rejecting human fetuses.

The relationship between personhood and humanity is important because both concepts and their relation have a strong say in determining who is included in care ethics. Personhood acts practically as a moral gatekeeper of anthropocentric approaches in the work of Kittay and Held. While Noddings declines establishing a "crystalized account of personhood" as a foundation of her theory, she does so because such a concept might be used to grant human infants and other animals the same moral value (Noddings 2013, 151–2). That is unacceptable to Noddings, who believes humans have a distinct moral status from other animals, let alone plants.

I am beginning my project here, because I think that it is a major hurdle to include plants in care ethics. If the theory assumes that only human beings can be cared-fors, then plants are left out of relationships included in care ethics. By examining the function that the assumptions of anthropocentrism have in these theories, I can better respond to how plants can be included in an ethic of care as cared-fors.

The Cared-for: Anthropocentric Accounts about Who (or What) Counts

For this project, I am assuming the carer in the caring relationship is a human moral agent. While I am not denying that non-human moral agent may exist, for the purposes of this project, it is easier to simply assume the carer is the standard moral agent—someone able to understand right and wrong in such a way that they are responsible for their actions. But who is the cared-for? For Noddings, the cared-for is a human. For Kittay and Held, the cared-for is a human person. In what follows, I examine each theorist's use of anthropocentrism to limit the domain of cared-fors to humans. While each push for a larger domain of cared-fors that includes human infants (often left out of some personhood theories), they ultimately fail to include non-human cared-fors that I argue should be included.

Noddings

Noddings' ethic of care centers on a caring relationship that is dependent, reciprocal, and dynamic (Noddings 2013, 86). The central relationship used in her analysis is that of a human mother and child (44). The human fetus, infant, child, and young adult is dependent on their human carer physically and emotionally. Despite medical advances, human bodies are still the primary incubation means for a human fetus to develop and eventually enter the world (Charest-Pekeski et al. 2021). And, of course, the work of motherhood takes a number of different forms: birthing, feeding, and raising a child, which can be done by different humans in a child's life (Ruddick 1989). A biological mother may give birth to an infant who is then raised by an adoptive parent, grandparent, or legal guardian. Regardless of the particular form, Noddings' emphasis is on the dependent, reciprocal, and dynamic quality of the relationship.

Highlighting the realm of human-interactivity and dependency, Noddings argues that an ethic of care exists within a dynamic relationship (Noddings 2013, 86). Dynamism allows for complex interaction and communication. The carer is attentive to the needs of the cared-for and able to change and adjust according with shifts in the cared-for. This dynamic relationship also has the ability to develop into something more complex and interactive over time. Part of this interactivity is connected to reciprocity. A reciprocal relationship resists exploitation, with the aim that each participant in the relationship is contributing to it. Rather than portraying the carer as a "Giving Tree"-esque (Silverstein 1964) character that sacrifices her very existence for the cared-for, the requirement of reciprocity ensures that the relationship is not entirely one-sided. While dependency can make relationships asymmetrical, Noddings' inclusion of the requirements of dynamism and reciprocity to ensure that exploitation is not the outcome of care.

Noddings' approach is especially distinctive in its anthropological approach. She has an account of natural care and ethical care. Natural care is a non-reflective practice that provides essential care to a cared-for and is prompted by feeling or impulse alone (Noddings 2013, 82). Ethical care is a moral practice that requires more complex interactions and works with feeling and reflection without using mere impulse to act out ethical care (86). Noddings associates natural care with other mammal mothers, and ethical care is something that we humans should do (83).

For Noddings, the requirement of reciprocity in care ethics eliminates the possibility of other animal and plant cared-fors. Noddings argues that other animals' development reaches a boundary beyond which human communication and interaction cannot pass, undermining the relationship's ability to be truly dynamic (Noddings 2013, 156). I am unsure why this type of dynamism matters for a moral theory and why Noddings so quickly dismisses dynamism within our relationships with non-humans. While the focus of this project is including plant cared-fors, it seems totally plausible that at least other animals could easily meet a threshold for dynamism and reciprocity in a caring relationship (Gruen 2015, 34–5).

Noddings contends that non-humans cannot reciprocate in a way that makes obligation morally required, as is the case of human relationships with other humans.

> We cannot refuse obligation in human affairs by merely refusing to enter relation; we are, by virtue of our mutual humanity, already and perpetually in potential relation. … In connection with animals, however, we may find it possible to refuse relation itself on the grounds of a species-specific impossibility of any form of reciprocity in caring.
>
> (Noddings 2013, 86)

For Noddings, identity as a member of a shared species binds us in obligation in a way that is more universal than the fragmented relationships humans form with other animals. They might be responsive to human given care, but reciprocity eludes our relationships with them.[1]

In discussing her interactions with her cat, Noddings emphasizes their relationship's connection with natural caring (Noddings 2013, 155). For Noddings, natural care comes about in the lives of both humans and other animals alike, while ethical care connects to morality in a distinct way, issuing obligations to care, rather than merely caring based on feeling or instinct (79–80). Ethical care is something only human carers can provide to human cared-fors. Other animal mothers and caregivers provide natural care, because it is born out of unreflective feeling and instinct. Ethical care is a distinctly human evolution of that instinct that expands to a moral obligation (156).[2]

Despite this clear preference for human cared-fors, Noddings does explain her reticence to rest her theory entirely on a conception of personhood: "We cannot accept an ethic that depends upon a definition of personhood if that definition diminishes our obligation to human infants. An ethic that forces us to classify human infants with rats and pigs is unsettling" (Noddings 2013, 151–2). As we'll see when addressing Kittay's work, there is a concern in care ethics about the "least of these" that other personhood theorists like Dennett omit from personhood as a concept: human infants and human cared-fors with varying cognitive capacities (120).

It is worth noting that while Noddings does not offer a particular definition of personhood, she both restricts the domain of care ethics to human beings and uses the words "human" and "person" interchangeably (Noddings 2013, 33, 82, 85–6, 127). So, while the term "person" occurs throughout her work, a specific definition is not made available. Her description of the relevant characteristics that are required for the cared-for—responsive (60), possessing a self (64), ability to recognize care (68), and the ability to reciprocate (74)—and restriction of care ethics to humans practically amounts to a requirement for carer and cared-for to be human persons, allowing that human infants—but not dogs or cats—qualify as persons.

I worry about Noddings' use of anthropological roots to differentiate human care from other creatures. Generic disciplinary territorialism aside, it may be useful to understand the form of care from an anthropological, developmental framework, but when we're talking about ethics, we need to examine any lingering issues that might be hanging on from the descriptive work. When Noddings points to the wide range of attitudes individual humans have towards different types of animals, she lets this be a stopping point for further argumentation about what attitudes could or *even should* be present. We may need to look to existing attitudes and examine what is possible, rather than what is actual, and revise our actual attitudes in light of changing context and moral reflection.

I also think there may be a concern that a universal obligation to care would require universal intervention. I take the most charitable interpretation of universal care not to be a mandate of a particular care to all cared-fors. Rather, ethical care is highly dependent on the needs and desires of the cared-for. It requires constant interpretation, inquiry, and reflection on the changing states of the cared-for. In some cases, intervention may be required, but not every caring relationship is as intensive as a human parent and infant child. And not every human must nurture a personal caring relationship with every other human.

Rather than mandating constant intervention, care ethics could acknowledge the difference between particular types of entities and modify the intervention of caring actions to each in accordance with their needs and the context of the relationship. For example, Noddings takes aim at a hermit monk when she argues that the monk has retreated from society and the meaningful relationships that one must have to participate in an ethic of care (Noddings 2013, 80). Some environmental ethicists would contend that the monk is meaningfully engaged in several relationships while on the mountaintop. His relationships with the land—soil, plant life, animal life, and geographical entities like mountains, streams, lakes, rivers, canyons, deserts, wetlands, etc.—are all morally significant. The moral obligations of the monk to these entities may be different than the obligations required from him were he to happen upon a human infant on his path. But dynamism and reciprocity are certainly present in the monk's relationship with other animals and other entities within his ecosystem. If the animal and plant cared-fors demonstrate reciprocity and dynamism in a morally relevant way, then they should be included as cared-fors. Chapter 2 will examine these capabilities in plants, which I argue qualify them in part as legitimate cared-fors in care ethics.

Kittay

Eva Feder Kittay characterizes ethical care as "a labor, an attitude, and a virtue" (Kittay 2019, 164). Care itself requires work on the part of the individuals involved and requires an attitude that asserts the importance of the well-being of the cared-for. The third factor, virtue, allows one to discriminate between good care and bad care; virtue imbues the quality and execution of care with moral value. This structure is somewhat different than Noddings, in that the distinction between care as an appropriate thing to do and care as an ethical theory is addressed in terms of virtue, a morally good character trait (168).

Kittay's view emphasizes the subject's experience in the ethic of care because care, as a virtue, is a golden mean between non-concern and paternalism. Avoiding absence and excess, care resists both total lack of concern and paternalism, a lack and excess of consideration (Kittay 2019, 208). Subjective buy-in of the cared-for to the work of the carer creates a way of determining whether care is successful. If the cared-for considers the actions of care to be unwanted, overburdensome, or side-stepping their needs and preferences, it will show up as paternalism. Kittay's approach is distinctive in her use of virtue ethics and in her elaborate discussions of cared-fors as human persons. She importantly expands personhood beyond human moral agents (a position that occasionally occurs in the personhood literature) arguing for the personhood of her daughter, Sesha, someone who the philosophical literature considers "on the margins."

Though Kittay uses the example of a plant as an object lesson throughout her discussion on the basic structure of care, she ultimately argues that non-humans are not cared-fors within care ethics because they lack subjectivity. Subjectivity, Kittay appears to claim, is a requirement for a person to be able to flourish (Kittay 2019, 199). The cared-for's flourishing is a requirement for care ethics. Without the cared-for's subjectivity, there is no non-objective way to ascertain whether the caring action is successful (191).

Apparently non-person living things can have a well-being: "We have a strong intuitive sense that a thing's or person's well-being must be positively affected by our actions if our actions are to be counted as care—at the very least, it ought not to be negatively affected" (Kittay 2019, 187). But Kittay reserves flourishing for persons (191).

While it is not a particularly controversial claim to deny the possibility of flourishing to plants, it is strange that plants can provide such a powerful example of the success conditions of care, only to be excluded from the ethic because they don't have subjectivity. At its core, flourishing is superseding your basic needs and using the excess energy to take up space and produce things in alignment with one's overarching goals or purpose. Because I am a human, my ways of flourishing are very different than my pup's, the brown thrasher's, or the jade plant's. But packing extra cognitive or sentience requirements to flourishing shows our profound lack of understanding of our fellow beings on this planet. They are useful for an object lesson, but not part of the ethic itself. Even in Aristotle's work, it was unclear that plants unequivocally qualify as the same types of living things that animals and humans are (Hall 2011, 32). Michael Marder and Matthew Hall, philosophers who focus on plant ethics, document such an in-between view of plants throughout human history (Hall 2011, Marder 2013).

I think this demonstrates a tension that is evident in Noddings' work as well. While there is intuitive pull in including other animals or plants within care ethics, both Noddings and Kittay explicitly want to avoid this. Noddings devotes an entire chapter to explaining why other animals, plants, and ideas do not belong in care ethics. Kittay's refusal to accept other animals in care ethics likely ties to a similar concern within Noddings worry about personhood—under some definitions, it excludes humans that they think should fit into care ethics or equates the moral value of humans and other animals. Kittay best articulates this concern in the following:

> If personhood is limited to those who possess certain intellectual capacities and to those who are productive, then my daughter would not be a person. But my daughter is a person. She is, after all, my daughter. How can she be anything but a person? If traditional conceptions of personhood are not capacious enough to include Sesha and those who share her impairments, we need a new definition"
>
> (Kittay and Feder 2002, 266)

This is a strong intuition and argument for a definition of personhood that includes humans who are "on the margins." I agree with Kittay that Sesha is a person.

While relationships are central to care ethics for Kittay, they also emerge as essential for understanding moral personhood. For Kittay, persons are comprised of and delineated by relationships that they have with others.

> We do not become a person without the engagement of other persons—their care, as well as their recognition of the uniqueness and the connectedness of our human agency, the distinctiveness of our particularly human relations to others and of the world we fashion"
>
> (Kittay and Feder 2002, 266)

These persons are ultimately only humans. It appears that Noddings is claiming engagement with other persons is distinctly connected to some essential activity of human beings and the human world. Of course, environmental ethicists who argue for the personhood of more-than-human beings would resist this conclusion.

Kittay's criticism of traditional personhood accounts—ones that exclude humans on the margins—illuminates both her definition of personhood and her resistance to include non-human entities as cared-fors. If an ethic of care should only concern traditional person cared-fors, it may not allow for two important phenomena in some women's lived experience that arguably should have a place in an ethic of care: having human dependents with various levels of cognitive capability and miscarriage (Lindemann 2002 and 2015). Some ethicists claim that all humans, regardless of cognitive capacity or fetal status, are persons. But the means by which theories admit these groups as persons is fraught with challenges (e.g., admitting the personhood of a fetus is sometimes the first step on a slippery slope to considering people who get abortions or have miscarriages as murderers).

Kittay responds to this concern using a personal example: her relationship with her daughter Sesha (Kittay 2019). Kittay asserts Sesha's personhood while rejecting traditional philosophical accounts:

> Because we as a family have been able to keep Sesha in our home and community, those who have made contact with her and have learned to see her as we who love her do have gained to perspectives on what it means to be a person. Seeing Sesha in her interactions with those who care for – and about – her reveals that being a person has little to do with rationality and everything to do with relationships – to our world and to those in it"
>
> (Kittay and Feder 2002, 266)

Kittay rejects the traditional notion of rationality as the determining factor in personhood. Rather, participation in relationships appears to be the ground floor for Sesha. I think this definition of personhood is an important revision of personhood. Though, I would argue that it is not merely relationships that are doing this work, but also Sesha's identity as a human being. In this theory, as a human, the other humans (i.e., human moral agents) that have relationships of care with her may set her apart as a moral person, while non-human animals, plants, and other animate and inanimate entities do not qualify. Kittay certainly resists the notion that Sesha is a moral person in the same way that other animals would be persons. She finds such a claim offensive (Gruen 2011, 72). For Kittay, the moral status of humans is distinct from other animals and entities. One way that oppression in human communities has functioned and continues to function is by denying members of a marginal society their full humanity and moral value. Instead, members of an oppressed group are considered less-than-human, and our language reflects this when derogatory terms from the non-human animal world are used to describe humans as a way of lessening the value of these humans. So, the legacy of equating humans considered on the margins or in other oppressed groups with animals considered lesser than humans is long. Because this oppression is still alive and well in several racist, misogynist, sexist, and ableist communities, there can be a push back against anything that resembles or has the possibility of facilitating dehumanization.

Regardless of this issue, being a part of a community and being in relationships is one alternative route to personhood. Sesha is a person and we can know that because the relationships that she has with other persons demonstrate and attest to her personhood. This is a truly radical shift away from rationality towards relationality. The source of personhood in the identification of personhood by others is slightly different from other relational accounts of personhood (e.g., ubuntu or Confucian).

Importantly for this project, some human persons also claim to identify and verify the personhood of plants. These plants may have relationships with other persons that constitute the plant's personhood. Also, human persons in relation with plants may consider these relationships to comprise their own identities. For example, some humans refer to themselves as plant parents. Not only core to how they see themselves, but the relationships with plants also cared-fors comprises who they are in their multi-species-community. This will be addressed in more detail within Chapter 2, but for now, it is valuable to flag the concern over Kittay's personhood definition embracing relational participation while rejecting non-human persons.

Held

Held's account considers care a value and a practice. This means that care is morally valuable in and of itself and should be sought out as an end in itself. Care is also a practice, meaning it must be acted upon and instantiated within the world. She establishes her core of care ethics as relationship, not a moral agent's character: "Virtue ethics focuses especially on the states of character of individuals, whereas the ethics of care concerns itself especially with caring *relations*. Caring relations have primary value" (Held 2005, 19). The relation is central: it may or may not include characteristics like reciprocity, progress, growth, etc. But given that care is a practice, an activity that requires a certain attitude and motivation, at least one of the entities involved should be a moral agent (42). A major distinguishing feature of Held's view is that care ethics are not merely a type of virtue ethics, but rather is a normative ethical theory distinct from all others.

Caring relations are at the center of Held's ethics of care. These caring relations are described as involving human persons within societies (Held 2005, 10). For Held, persons are fundamentally relational beings: "The ethics of care, in contrast, characteristically sees persons as relational and interdependent, morally and epistemologically" (13). This sense of self ties directly into feminist philosophical attempts to veer away from an identity built upon the ideal of a rational, disconnected individual, and lean into a fundamentally relational being (13).

The connection between selfhood and personhood continues the legacy of coupling human identity with moral status. Despite briefly mentioning caring relations with other persons and their environment (Held 2005, 159), Held makes clear that an ethic of care exists in relations between human persons, as she seems to use "person" and "human" interchangeably (10).

By highlighting how these theories justify the use of ethical care only among humans and human persons, I bring to light the assumption of anthropocentrism. While Kittay extends personhood to marginal human cases, she stops short of extending personhood to other animals and other entities. For an ethics of care to meaningfully work outside of the human community, this relationship between anthropocentrism and their ethical structure must be addressed. With properly applied pressure, I argue that an ethic of care can, and should, include other non-human and non-person entities.

The Function of Anthropocentrism in the Cared-for

Anthropocentrism and personhood function in different ways within the work of Noddings, Kittay, and Held. Importantly for my project, humanity and personhood identify who can count as a cared-for in care ethics. Kittay and Held are crystal clear in their use of person cared-fors. The focus on the caring agent in Kittay's project, along with her work on the labors of love in marginal cases, show that personhood is central to her ethics of care. And personhood, for her, is a community made of exclusively of human beings. Importantly, more human beings that in traditional personhood theories that exclude some folks with a range of disabilities, but human beings only nonetheless. Taking a slightly different approach by making relations rather than virtue central, Held explains that caring relationships are between persons. Relationships between persons and non-persons are not party to ethical care. Rejecting definitions of personhood that require moral agent threshold levels of rationality, care ethics is precisely interested in cared-fors like infant human beings and human beings with different cognitive capabilities. Kittay and Lindemann realize the necessity of expanding the traditional ratiocentric accounts of moral personhood for a complete ethic of care. But they do not expand it enough, in my view.

Quizzically, Noddings admits that some animals could be considered cared-fors. However, species differences (and members of some species, like rats, she considers a threat rather than potential subject-of-care) prohibit a universal obligation to care for other animals. Because humans share species membership with other humans, we have a universal moral obligation to care for each other (should the opportunity or demand arise). But other animals are more of a case-by-case potential.

Most clearly in these accounts, personhood appears as an indicator of moral status reserved for humans revealing a type of anthropocentrism. Noddings and Kittay explicitly express concern over granting equal moral status to human infants and other animals. Noddings' reticence to use a specific definition of personhood that could be used to equate the moral value of humans and non-humans demonstrates this concern. These examples of care ethics apply to relationships between humans because human beings have a special moral status that elevates them apart from other non-human entities. In many circles, this anthropocentrism is intuitive or considered common sense. So, it is not surprising, given the philosophical history of the concept of personhood in Western philosophy, that this would be part of personhood's function when it appears in care ethics.

Perhaps this stark setting apart is justified in other contexts, but in care ethics, I am unconvinced that it is necessary. Care ethics centers relationships between carers and cared-fors. Many humans with companion animals would be able to identify themselves as carers and cared-fors. The slogan "Who Rescued Who?" from the rescue animal sector might even suggest that non-human animals can return this care. "Plant parents" might also see ethical care in their relationships with their plant cared-fors. Admitting non-humans as cared-fors may yet allow for having different moral statuses for various cared-fors. Given arguments for non-instrumental moral value for other animals and plants that will be discussed in Chapter 2, denying them the status of cared-for because they are not humans is simply not good enough.

Anthropocentrism and personhood also function as a stand-in for a collection of characteristics in the cared-for that are relevant for care. What makes someone or something a proper cared-for? They have moral status, they depend on the carer in some way, they respond to actions that affect them, and they can change and require context-dependent responses from their carers to ensure that the world goes well for them. As useful as personhood could be in bringing these elements together, if it obfuscates who or what can qualify as a cared-for, enumerating the bundle of characteristics might be better. I think in this case, the legacy of "person" and "human" might leave out legitimate cared-fors. In the most uncontroversial case, companion animals seem to clearly count as cared-fors. In the more controversial case that I'll be arguing for, plants should too. In these cases, claiming that the cared-for must be a human person is leaving out legitimate cared-fors, and should be rejected.

Conclusion: Suspending Anthropocentric Assumptions and Making Space for Plant Cared-fors

In each of the aforementioned accounts, ethics of care employs anthropocentrism. Noddings provides some argument as to why animals, plants, and other things should not qualify as cared-fors in an ethics of care. Though, under scrutiny, I am doubtful that her arguments about other animals would hold up against contemporary scientific behavioral research on the capabilities of some other animals. Instead of stopping here, and simply asking Noddings, Kittay, and Held to justify the exemption of other animals, plants, and other living organism cared-fors, I interrogate what I consider to be an underlying issue in each of the theories above: the anthropocentrism. In my view, the limiting of cared-fors to humans poorly justified.

A larger issue underlying this anthropocentrism in care ethics is the concern that arguments justifying the exclusion of non-humans are unnecessary because care is distinctly human. Care emerges as an ethically valuable concept from observations of human life. Humans depend on others for their survival. The dependency that Noddings, Kittay, and Held are captivated by springs out of human social lives and relationships. At first blush, plants, let alone other animals, do not even occur within the context of human dependency on care. When human infants emerge into the world, they largely depend on other humans for their survival. Prior to being born, the fetus is physically dependent on the mother's body. Even after viability, it is impossible for the viable fetus to survive on its own, without the intervention of other human care givers. When we face illness or aging or disability, we rely on other humans for our care. Some might argue that in less drastic circumstances, we are constantly receiving and giving care, both for survival and thriving. So, an ethic of care emerges from these fundamental caring relationships that concern our very survival, development, and quality of life.

This locus of dependency on human relationships and focus on affect in prompting or facilitating care might explain an initial disinterest in looking beyond these human communities for instances of ethical care. In the Western canon, ethical theories regularly prioritize human beings to the point of exempting other entities from ethics, outside of instrumentality (Hall 2011, 164–5). Under these theories, non-human beings can be used, but not have value in their own right; they are too different from us.

However, in the wake of environmental ethics and subsequent criticisms of Western philosophy's general anthropocentrist starting point, it becomes unclear why other entities outside of the human species cannot qualify as proper cared-fors. An initial consideration that care ethics extends beyond human communities is reasonable. Human lives are ephemeral, fragile things. We are dependent on other non-human entities for so much that we require in life. We could simply attribute this to the instrumental tradition: that non-human entities are simply valuable to us, morally speaking, because they can provide some service or meet some need. Even outside the realm of human-and-other-animal relationships, humans depend on plant life for our physical survival. We rely on at least grains (rice, wheat, corn, etc.), beans and legumes, a variety of vegetables and fruit. Surely some humans also consume animal products (flesh, eggs, milk, bones, skin, etc.) to survive physically. We are dependent, in the strictest sense, on plants and other animals for our physical survival, and it is worth considering the possibility of care in these relationships.

But care ethics goes beyond this type of dependency. We are not only dependent on other humans, alongside other animals and plants, for physical survival. Our social communities are comprised of different beings. We work with other animals and plants to thrive and flourish. Relational connections of care, rooted in survival, blossom into more intricate interactions of well-being. If anthropocentrism cannot account for these complex and multi-species relationships of care, perhaps it should be rejected.

In the following chapter, I embark on the challenge of including plant cared-fors in care ethics. I propose five potential options:

a consider plants to be persons,
b consider plants to be subjects-of-a-life,
c consider plants to be valuable as living beings,
d consider plants to be valuable because they can flourish, or
e consider plants as kin.

The first option keeps personhood a requirement for the cared for, while opening it up beyond humans. The second option takes Tom Regan's subject-of-a-life criterion to emphasize views that require sentience for cared-fors. The more radical options (c), (d), and (e) reject both anthropocentrism and sentiocentrism to offer an inclusive care ethic that can include conservative accounts of plant-life as cared-fors. I aim to provide defenses for each of these positions in the literature while resisting endorsing a particular view. So, I argue that anthropocentrism should be dropped in favor of a care ethic that can include more-than-human cared-fors.

Notes

1 Companion animals can express their appreciation of care, as many pet parents are familiar with, but this falls short of reciprocity.

2 Noddings may have an obligation to her cat that looks like ethical care because they have established a relationship. But she claims that these instances of care are more particular to individuals and cannot be assumed to apply more broadly because humans hold different attitudes towards various other animals. Also,

different species of animals (in her example, rats) do not require a care response. As I read this section of Noddings, I could not help but think of my cousins who have pet rats. This differentiation amongst species seems to be relying on personal intuitions. But perhaps that is the point—the broad range of attitudes and intuitions about other animals means that there will be differing views on which we could have obligations of care for and ones for which we are not required to care.

References

Batavia, Chelsea and Michael Paul Nelson. "For Goodness Sake! What Is Intrinsic Value and Why Should We Care?" *Biological Conservation*, 209 (2017), 366–376.

Bekoff, Marc. *The Emotional Lives of Animals: A Leading Scientist Explores Animal Joy, Sorrow, And Empathy—And Why They Matter*. New World Library, 2007.

Callicott, J. Baird. *Beyond the Land Ethic: More Essays in Environmental Philosophy*. SUNY Press, 1999.

Card, Claudia. "Review: Caring and Evil." *Hypatia*, 5(1) (1990), 101–108.

Charest-Pekeski, Alex J., Ayman Sheta, Luiza Taniguchi, Mark J. McVey, Alejandro Floh, Liqun Sun, Tanroop Aujla, Steven K. S. Cho, Jiaqi Ren, Lynn Crawford-Lean, Celeste Foreman, Jessie Mei Lim, Brahmdeep S. Saini, Marvin Estrada, Anson Lam, Jaques Belik, Dariusz Mroczek, Megan Quinn, Stacey L. Holman, Jack R. T. Darby, Mike Seed, Janna L. Morrison, and Christoph Haller. "Achieving Sustained Extrauterine Life: Challenges of an Artificial Placenta in Fetal Pigs as a Model of the Preterm Human Fetus." *Physiological Reports*, 9(5) (2021), e14742.

Curtain, Deane. "Toward an Ecological Ethic of Care." *Hypatia*, 6(1) (1991), 60–74.

Dennett, D.C. *Brainstorms: Philosophical Essays on Mind and Psychology*. MIT Press, 2017.

De Waal, Frans. *Are We Smart Enough to Know How Smart Animals Are?* W. W. Norton & Company, 2016.

Di Paola, Marcello. *Ethics and Politics of the Built Environment*. Springer, 2018.

Donovan, Josephine and Carol J. Adams (eds). *The Feminist Care Tradition in Animal Ethics*. Columbia University Press, 2007.

Fesmire, Steven. "Pragmatist Ethics and Climate Change." In *Moral Theory and Climate Change*. Routledge, 2020.

Gilligan, Carol. *In a Different Voice: Psychological Theory and Women's Development*. Harvard University Press, 1982.

Goodpaster, Kenneth E. "On Being Morally Considerable." *The Journal of Philosophy*, 75(6) (1978), 308–325.

Gruen, Lori. *Ethics and Animals: An Introduction*. Cambridge University Press, 2011.

Gruen, Lori. *Entangled Empathy: An Alternative Ethic for our Relationships with Animals*. Lantern Books, 2015.

Hall, Matthew. *Plants as Persons: A Philosophical Botany*. State University of New York Press, 2011.

Held, Virginia. *The Ethics of Care: Personal, Political, and Global*. Oxford University Press, 2005.

Hoagland, Sarah., "Review: Some Concerns about Nel Noddings' 'Caring'." *Hypatia*, 5(1) (1990), 109–114.

Hourdequin, Marion and David B. Wong. "A Relational Approach to Environmental Ethics." *Journal of Chinese Philosophy*, 32(1) (2005), 21.

Hourdequin, Marion. *Environmental Ethics: From Theory to Practice.* Bloomsbury Publishing, 2015.

Kant, Immanuel. *Groundwork for the Metaphysics of Morals.* Translated by Allen W. Wood. Yale University Press, 2002.

Katz, Eric and Andrew Light (eds). *Environmental Pragmatism.* Routledge, 1996.

Kittay, Eva Feder and Ellen K. Feder (eds). *The Subject of Care: Feminist Perspectives on Dependency.* Rowman & Littlefield Publishers, 2002.

Kittay, Eva Feder. *Learning from My Daughter: Valuing Disabled Minds and Caring That Matters.* Oxford University Press, 2019.

Kopnina, Helen, Haydn Washington, Bron Taylor, and John Piccolo. "Anthropocentrism: More than Just a Misunderstood Problem." *The International Journal of Ecopsychology (IJE)*, 3(1) (2021).

Lindemann, Hilde. "Miscarriage and the Stories We Live By." *Journal of Social Philosophy*, 46(1) (2015): 80–90.

Marder, Michael. *Plant-Thinking: A Philosophy of Vegetal Life.* Columbia University Press, 2013.

McShane, Katie. "Why Environmental Ethics Shouldn't Give Up on Intrinsic Value." *Environmental Ethics*, 29(1) (2007), 43–61.

Molefe, Motsamai. *African Personhood and Applied Ethics.* NISC [Pty] Ltd, 2020.

Nelson, Hilde Lindemann. "What Child Is This?" *Hastings Center Report*, 32(6) (2002), 29–38.

Noddings, Nel. *Caring: A Relational Approach to Ethics & Moral Education*, 2nd edn. University of California Press, 2013.

Puig de la Bellacasa, María. *Matters of Care: Speculative Ethics in More than Human Worlds.* University of Minnesota, 2017.

Regan, Tom. *The Case for Animal Rights.* University of California Press, 1983.

Rolston, Holmes III. *Environmental Ethics: Duties to and Values in the Natural World.* Temple University Press, 1988.

Ruddick, Sara. *Maternal Thinking: Towards a Politics of Peace.* The Women's Press, 1989.

Schechtman, Marya. *Staying Alive: Personal Identity, Practical Concerns, and the Unity of a Life.* Oxford University Press, 2014.

Siderits, Mark. "Persons and Selves in Buddhist Philosophy." In Antonia LoLordo (ed.), *Persons: A History* (pp. 301–325). Oxford University Press, 2019.

Siderits, Mark, Evan Thompson, and Dan Zahavi (eds). *Self, No Self? Perspectives from Analytical, Phenomenological, and Indian Traditions.* Oxford University Press, 2011.

Silverstein, Shel. *The Giving Tree.* Harper Collins Publishers, 1964.

Taylor, Paul W. "The Ethics of Respect for Nature." *Environmental Ethics*, 3(3) (1981), 197–218.

Tronto, Joan. *Moral Boundaries: A Political Argument for an Ethic of Care.* Routledge, 1993.

Warren, Mary Anne. "On the Moral and Legal Status of Abortion." *The Monist*, 57 (4) (1973), 53–54.

2 Vegetal Matters

Plant Cared-fors in Care Ethics

Who or What Matters: Moral Value and More-Than-Human Cared-fors

In considering how a plant could be a cared-for within the framework of care ethics, I think the two major issues that one must face are establishing the plant's moral status and determining whether the plant has the capabilities needed to participate in an ethically caring relationship. I want to quickly flag that this might ring as a conservative approach to some. Within environmental ethics as a discourse, questions of moral status—especially who or what has intrinsic moral value—go in and out of vogue. For this project, I consider myself to be building a bridge to meet people where they are in their views about intrinsic moral value. My approach could be considered quite conservative, reluctant to automatically grant to plants capabilities that others in critical plant studies readily assign to plants (e.g., consciousness, personhood, etc.). What one is willing to grant that a plant can do or be will inform whether that person thinks plants can qualify as cared-fors. And there are radical options on the spectrum of beliefs that humans have about plant capabilities. Rather than arbitrate the truth of each category that plants could fall under that grants them moral status, I want to look at five possible sources of that moral value and provide an introduction to the parts of those views that might be more or less difficult for someone to accept. In doing this, I aim to center the importance of moral status as one tool for making sense of who or what properly qualifies as a participant in the ethically caring relationship. I want to provide space for pragmatic value pluralism, allowing for the wide and unsettled accounts of what plants are and what they can do, both in the sciences and humanities.

As I consider each framework as a potential for qualifying plants as having the proper moral status required by cared-fors in care ethics. People who are skeptical that plants are the types of beings that could participate in care ethics as a cared-for will hopefully benefit from this careful multi-option approach. Folks who are already on board with plants being moral persons might find this process tedious. But as an analytic philosopher, it is essential to pull apart these threads of beliefs about and descriptions of plants in order to ascertain what plants could be and how they could fit into care ethics.

DOI: 10.4324/9781003603665-3

To consider other animals and plants in care ethics, anthropocentric requirements of cared-fors must be confronted and alternatives must be explored. To accomplish this, a good account of cared-fors should include both inherent value and care-relevant capabilities for the cared-for. Providing an account of inherent moral value for the cared-for can help elevate the relationship between carer and cared-for to domain of ethical care. But the ethically caring relationship also includes consideration of certain capabilities that make care possible in the relationship. To meet that end in this chapter, I provide an array of accounts in animal ethics, environmental ethics, and plant ethics literature that respond to non-human cared-fors.

I consider five possible options for cared-fors that could include plants:

a use existing personhood accounts to include plants as persons (i.e., argue that plants have characteristics associated with anthropocentric personhood, including reason, autonomy, agency, etc.);
b use Tom Regan's subject-of-a-life to incorporate plant cared-fors as moral patients;
c use biocentric accounts of moral value to establish plants as intrinsically valuable;
d use the concept of flourishing to establish plants as intrinsically valuable; or
e use the concept of kinship to establish plants as intrinsically valuable.

I offer a literature-review laden account of each of the above approaches, evaluating the merits of each.

Option A: Plant Persons

One clear way of including plants in care ethics is to provide an account of moral personhood for plants. If a care ethicist merely requires the caring relationship to be between persons, then giving an acceptable account of plant personhood could answer that concern. "Merely" is a slippery word though because establishing an acceptable account of plant personhood in our current cultural climate and constellation of scientific paradigms is a complicated, controversial, and challenging task that points to the importance of one's worldview and cultural context in determining what is possible and what is true. A rich literature exists in animal ethics that supports considering non-human animal persons (e.g., Warren 1997, Varner 2012, Hutton 2019). Plant personhood, markedly more controversial than animal personhood, is also a part of a growing, interdisciplinary literature (Hall 2011).

As discussed in Chapter 1, and nicely summarized here by Annette Baier, personhood as a philosophical concept can be fraught.

> The schizophrenia of modern philosophy of mind and of the person, its back-handed tribute to our mammalian nature, to our place in a sequence of generations of living, mating and dying human persons, is

> our joint responsibility… The term “person” is a status term, and it is *our* term. It is we who have to decide what that status is and whether we give it to a human fetus, other animals, to corporations.
>
> (Baier 1991, 12–13)

I appreciate Baier’s depiction of the complicated history and existence of personhood as a concept in philosophy. Personhood has been a source of contention, leading to some humans to be labeled non-persons, or lacking moral status. The historical and contemporary weaponization of personhood makes one wonder if it is even worth the cumbersome project of utilizing the concept in contemporary work. As Baier writes,

> Sometimes one wishes we could just drop the term, so heavily freighted is it with its theological past and with current controversies. But words are hard to kill. The only realistic strategy is to make do with the old concept, heavily burdened though it is.
>
> (Baier 1991, 12–13)

While personhood might have traces of “our mammalian nature,” it is still “*our* term” that we can change through restriction or expansion. And so, in this spirit, the first approach to including plant life as cared-fors in care ethics is to consider plants as persons.

There are many accounts of moral personhood. Most of them include a collection of capabilities (Palmer 2010, 44) that are historically associated with humans. Following Kantian moral theory, personhood is most commonly associated with autonomy and reason (Gruen 2011, 57). Entities that have moral status are ones with a set of capabilities that justify their moral considerability. Reason is such a capability, because it enables autonomy, a type of freedom that allows one to make moral choices and participate in moral reasoning. Autonomy is understood to be morally valuable and worth honoring because of this freedom and subsequent responsibility. As mentioned before, only some human beings have reason and autonomy under many of these historical and contemporary personhood accounts.

Matthew Hall (2011) highlights autonomy in his case for plant personhood. Hall’s argument for plant personhood rests heavily on his ability to defend the following:

> On the basis that plants clearly demonstrate self-directed growth, the communicative, sentient, intelligent nature of plants is also established. This is the recognition of plants as other-than-human persons – a powerful way of incorporating plants within social and moral relationships of care and nurturing.
>
> (Hall 2011, 161)

Self-direction is akin, for Hall, to autonomy. And, following Kant, autonomy is a capability that warrants protection and grants holders moral considerability.

Hall's analysis engages historical and contemporary religious and cultural attitudes about plant value, to show that there are some human groups that start from starkly different assumptions about plant capabilities (Hall 2011, 4–6). While Abrahamic monotheism and disenchanted scientific atheism might dominate the cultural climate globally at the moment, the acknowledgment of multiple religious and cultural human groups that grant moral status to at least some plants is necessary in a discussion of plant personhood. Some discourses begin with plant personhood being the rule, rather than an exception. These include particular strains of belief systems within Hinduism, Jainism, Buddhism, indigenous animism, and paganism (10–11). This belief in plant personhood does not simply get plucked out of the ether; it has an established presence in different belief systems, both historic and contemporary.

Of course, some humans believing a thing does not make it true; but it does begin to push back against the initial incredulity of an unsympathetic audience. By dismantling the (in Hall's view, artificial) gulf between humans and other things in the world and getting his interlocutor to grant that plants are not a wholly different ontological type than humans, he can convince them, using scientific advances in botany, that there is only a difference in degree, not kind, between humans and plants. If the difference is only in degree, then plants could have the capacities invoked in personhood: autonomy, communication, sentience, and intelligence. If plants have those capacities, then they should count as moral persons. This follows Val Plumwood's ecofeminist concerns about patriarchal and androcentric dichotomies and the dividing of natural entities into types. Plumwood (2002) argues that these dichotomies, rich with corresponding value hierarchies of good and bad, lesser and greater, have built overtime an unjustifiable difference between one and the other. For example, common dichotomies may include culture and nature, men and women, and humans and other animals. In each case, the former is assumed either have a higher degree of value or be the standard against which the latter is measured. Hall is picking up Plumwood's challenge to create an environmental ethic of difference in degree, rather than in kind (Hall 2011, 137).

The building blocks for a contemporary account of plant personhood from the sciences begins, for Hall, with Darwin and evolutionary theory. The scientific paradigm of Darwinian evolution provides more fertile ground for discussing the biological relatedness of humans and plants. The positing of common biological ancestors between existing humans and plants begins breaking down the case for total othering and human exceptionalism. Perhaps there is not as stark of a contrast between plant life and human life as we have cultivated in Anglo-European philosophy; our relatedness is more primary than our difference.

If the reader is convinced of Hall's claims of kinship (i.e., that given our biological and historical relatedness, differences in capacities are differences in degree not kind) then we may simply need to apply existing methods for determining personhood to different plants. Referencing the work of Anthony Trewavas (2014), a plant physiologist known for his work on plant behavior, and *Communication in Plants: Neuronal Aspects of Plant Life* edited by František Baluška, Stefano Mancuso, Dieter Volkmann (2006), Hall connects scientific capacities with the ones required for moral personhood. Hall's account of personhood requires autonomy, communication, sentience, and intelligence.

While this chapter is not strictly a scientific investigation into the claims of scientists like Trewavas, the reception of their claims within the contemporary plant biology community is worth acknowledging. This is true especially for capacity-dependent accounts of cared-fors. If personhood is determined through a composite of capacities, the scientific support for such capacities is essential to making their case.

Trawavas' work emerges from the reigning paradigm of plant biology even as it challenges some contemporary conclusions and frameworks. A fellow of the Royal Society trained in molecular biology and earning his doctoral degree in Biochemistry from University College London and holding the position of Professor in Plant Biochemistry at the University of Edinburgh until his retirement in 2004 (The Royal Society and Science Museum Group). Contemporary paradigms of plant biology view plants as a diverse set of living beings in the Plantae kingdom. Many are identified as plants in part by their ability to photosynthesize (i.e., absorb solar energy from the sun and turn them into energy for the plant body). These beings respond to their environment similarly to other living beings (e.g., bacteria and animals). So, it is reasonable to consider that his utilization of the tools and practices of this scientific paradigm can justify his theoretical claims. The scientific critical process on the subject of plant behavior and plant intelligence is likely still in its early days. As Marc Bekoff, a prominent animal biologist and behavioral expert, mentioned in a recent interview, the field of critical plant studies appears to be at a stage that consciousness studies in other animals were 200 years ago (Bekoff 2022).

It is worth flagging that this struggle over describing and defining plant capabilities is not a new area of contention. While the proposal of naming some scientific approaches "plant neurobiology" is a more recent controversy, the skepticism around plant capabilities and their descriptions has persisted at least since Peter Tompkins and Christopher Birds' *The Secret Life of Plants* (Pitt 2025). A major issue within contemporary Critical Plant Studies and the plant humanities is just how similar these functions are to other living beings, especially humans, and our use of language to describe them. One of the core issues here is over the meaning of sentience and consciousness, an infamously challenging philosophical concept. Theorists (both in the sciences and other disciplines, including the humanities) wrestle over how to describe the activities of plants. When one uses language that is fairly similar to the activities or capabilities of humans, charges of anthropomorphism can arise.

Anthropomorphism is many things. Considered a negative charge in much recent work examining more-than-human communities, it can describe the work of people working with non-human beings that projects human capabilities. So, if my dog seems to enjoy the company of another dog evident in his behavior (e.g., tail wagging, play bowing, whining when he sees the dog across the street, sniffing the dog's messages written in urine all over the neighborhood), and I call the other dog his friend, some might cringe. Others would not bat an eye. Whether the description of their relationship as a friendship is acceptable depends on one's view of what dogs are capable of. Can my dog have a friendship? In human-centered definitions of friendship, he probably cannot. Because friendship may demand a high bar built on rigorous intellectual, spiritual, social, and emotional engagement associated with (and presumably only accessible to) humans. However, if friendship is, by definition, a social relationship of preference, trust, and comfort, it may be open to social animals that are able to communicate with one another. Under this definition, my dog can likely have friendships (and possibly also enemies). So, whether I am unjustifiability projecting the practice of friendship to a non-human animal might rest on contention over whether this is a distinctly human activity. Am I taking a human phenomenon and forcing non-humans into its form? If so, this could be a form of pernicious anthropomorphism. And it could lead to a number of problems, like holding my dog accountable for things that he shouldn't be punished for because he is not a human being. Also, it could be considered a negative outgrowth of anthropocentrism; if all other beings must be judged according to human traits, capacities, and rules.

However, there are also other ways to understand anthropomorphism. One could look at the tendency of humans to use human experience or concepts to project onto the non-human world as either non-perniciously natural or beneficial for understanding the world. I'll be honest—I think my dog is capable of having both friendships and enemies. But I think this has more to do with his identity as a social being. My description of those social relationships in those terms (i.e., friendship) has more to do with my identity as a social being. So, I pick out things I find valuable or resonant in the more-than-human world. In some cases, this can be a good thing for ethics, I think. For example, the anthropomorphic impulse to ascribe friendships to my dog might unconsciously also support my view that he has a life of his own, independent of me, the human, that is valuable in its own right. By projecting my experience onto his, and understanding what is true and not true about that projection, I can better understand him on his terms. However, that initial projection created a bridge to his experience and enabled a type of empathy to connect with him.

In the context of plants, these debates about anthropomorphism and how to describe the capabilities of plants are ongoing. Recently, there was a helpful exchange between Trewavas, Paco Calvo, and Lincoln Taiz in *Protoplasma*. Each has a different view of what plant intelligence means, and as Calvo notes, this is an important part of navigating and renegotiating professional disagreement (Calvo, Raja, and Segundo-Ortin 2025). Trewavas and Taiz have been in a perennial struggle over the controversial area of plant neurobiology. Trewavas is supportive of using terms like intelligence and behavior to describe plant activity, while Taiz is critical (Trewavas 2025, Kingsland and Taiz 2025). Calvo is a philosopher and works in the space of cognition and consciousness, as well as the founder of the Minimal Intelligence Lab in Mercia, Spain. And so, Calvo enters the conversation primarily as a philosopher (evident in his discussion of fallacious vs non-fallacious exchange) and an interdisciplinary scholar working in the space of plant intelligence. In this three-party exchange, disagreements involve the proper and improper use of historical sources for evidence (including theorists used here like Charles Darwin and Barbara McClintock) and incorrect quotes used to leverage specific interpretations. As Solvejg Nitzke argues in *Making Kin with Trees* (Nitzke 2025), these areas of difference and uncertainty can create new spaces for research and prompt better scientific and non-scientific questions to pursue in the uncertainty and empty spaces of not-knowing-yet.

At this point, I think it is reasonable to side with Trewavas or the establishment, represented by Taiz, or the more nuanced position forwarded by Calvo and the Minimal Intelligence Lab. There are growing numbers of scholars in different disciplines holding up Trewavas' work and following in his investigative trajectory. As someone entrenched in interdisciplinary academic and professional plant studies networks, I tend to err on the side of Trewavas. Skeptics of my view might err on the side of the establishment. So, I worry that this is an impasse for some. I am also not confident that I can convince a skeptic that plants are rational and autonomous in the philosophical sense, meaning in a similar enough way to humans as to qualify them for moral personhood. If one were to agree with the work of Trewavas that plants behave, this may still leave a space between the ability to act and the ability to reason. Still, I think it important to go through relevant factors to personhood to ascertain the coherency and strength of these accounts, even if they won't (or can't) entertain the claims.

Autonomy

Providing accounts of plant autonomy could fill an entire book itself, so I won't be providing a complete overview of all the possible meanings of plant autonomy here. But I take Hall to be pursuing the morally significant core of autonomy that sets persons apart from other beings: namely, they make their own decisions that result in the ability to take self-directed action. For some approaches, this decision-making and enacting capacity indicates an ability to implement one's reason in the world. Trewavas (2014, 85) offers a discussion of plant choice and change in the case of two carnivorous plants: *Dionaea muscipula* (Venus flytrap) and *Drosera rotundifolia* (round leaved sundew). Carnivorous plants are a favorite subject for scientists discussing plant behavior because their motions can be quick and easily observable on a human sensory scale. When something lands on the gooey opening of the *Dionaea muscipula* or touches the tentacles of the *Drosera rotundifolia*, parts of the plant react quickly to the stimulus. The plants identify whether the object is an edible morsel or a nonedible thing and begin digesting the former or opening back up in response to the latter. Trewavas utilizes this example as a plant making the choice to move forward with digestion or deciding that it will not use energy to digest something. Other examples of plant choice include grasses choosing bare land over covered, pea plants choosing uninhabited soil, and vines choosing a suitable tree to scale (84).

These instances of choice illustrate a plant's ability to evaluate different options and move itself towards one option rather than the other. In some cases that means closing its trap, extending its roots, or circling its tendrils. These examples demonstrate autonomy because the plant is evaluating information in its environment in order to choose a specific action. As I'll address below, skeptics might consider this account of autonomy as different in kind rather than degree when considering the autonomy ascribed to human persons. But for the time being, Hall considers choices and decisions like this made by the plant an enactment of autonomy on a continuum with human choice and decision.

Communication

Plant communication appears quite different than methods of human communication we use to interact with each other and other animals (Thomas and Parks 2025, 17–18). Rather than focusing on vocalizations and rapid bodily movements, the literature on plant communication emphasizes chemical signaling. Trewavas notes that plant chemical signaling can occur when a plant is set upon by an outside threat (Trewavas 2014, 185–6, 268). These chemicals can be directed at the organism attacking the plant, letting them know that the plant is not suitable food. The chemicals can also be released in a way that is picked up by the same species to alert them to the threat. Trewavas contends that competition between plants also evinces communication. For example, there are spaces between certain desert plants that fight for limited resources (Trewavas 2014, 170, 188). These spaces present communication between plants because of the chemical signaling that the roots of plant A express "stay away" to the roots of plant B. Other plants may have cooperative and interactive relationships between roots, like the trees that feed sugar to keep trunks of fallen trees alive (Bader and Leuzinger 2019). Root systems are a major hub of communication. Not only do the roots of one plant communicate with each other, they communicate through mycorrhizal networks with fungi, bacteria, and other plants (Trewavas 2014, 133, 173).

This communication clearly is not through animal-like vocalization or timed movements, but rather through chemical signaling, root growth, and plant movement. Again, skeptics might have concerns—in part because human language has long been referenced as a morally relevant factor differentiating humans from other animals, let alone plants (Regan 1983, 38–9). But if we are defining communication as interactions that share information, then perhaps this activity can count given that the plant might share its state or experience with others in the ecosystem.

Sentience (and Self-Awareness)

Sentience here may include experiencing pleasure and pain but also includes awareness of other environmental and social forces. Accessing plant pain and pleasure sensations may be a black box of sorts; acknowledging consciousness in other animals seems to rely on human recognition of other animal brains and resonance with the behavior of other animals (movements and vocalizations). But Trewavas and other plant neurobiologists consider the activities of plants to be behavior. Despite plants not having brains and not moving or vocalizing as similarly to us humans as other animals do. Does that mean that plants are not sentient? Trewavas and Hall would urge us to resist drawing that conclusion.

Trewavas contends that all living organisms are conscious in a higher-order sense: "Higher order consciousness starts with perception through the sense organs and then control of muscular activities that drives the response. Perception, assessment, response—these are fundamental behaviours for all organisms" (Trewavas 2014, 256). Plants have sense organs to receive information from the environment and respond to light, soil qualities, and the position of other plants and animals. Plants exhibit their ability to be aware of their environment in activities like growing further into spaces while avoiding others and moving in relation to the position of the sun and amount of solar access.

A second important type of awareness for personhood accounts is self-awareness. In plants, there is an awareness of the plant itself as opposed to other bodies. This is observable not only in root behavior, but also in the communication patterns of pea pods that were genetically identical but separated in different ways—one pair grown separately, while another split from the same seedling (Trewavas 2014, 188). The relationship between self and personhood is fraught in the philosophical literature. Returning to the approach mapped out in Chapter 1, examining the function of personhood is fruitful for determining what is necessary.

Intelligence: Ability to Reason and Problem-Solving

Intelligence is a difficult concept in this arena and has many definitions (Calvo and Lawrence 2023). Intelligence in plants does not mean the solving of arithmetic, but I would hope it also does not mean that in the ethical personhood context. If one's personhood depended on their ability to count, I worry that personhood would rise or fall on capabilities that are not morally relevant. Definitions of intelligence inherited from Kant likely focus on the ability to reason. Evaluating whether plants are able to access reason in the way that Kant argues humans do is likely a fruitless endeavor because it is so bound up in human identity. However, if intelligence and reason have more practical value, as in an ability to problem-solve for the survival of an organism, then Trewavas has evidence of plants doing this. One example of problem-solving involves the participation of plants in decision making that Trewavas argues is not unlike problem-solving approaches identified in game theory (Trewavas 2014, 178). For example, there are a limited amount of nutrients in any given environment. Plants use signaling to providing information about the plant's decision to cooperate with other network organisms (through mycorrhizal networks) or to compete with them. These decisions require plants to have some memory of past interactions and require the synthesis of multiple sensory inputs. This is intelligence.

Beyond Hall's *Plants as Persons*, examples of theorists engaging plant personhood are growing in the plant humanities and critical plant studies spaces. For example, in *Storying Plant Communication*, Mariko Thomas and Melissa Parks provide a positive analysis of pedagogy that includes indigenous and animist attributions of personhood to more-than-human entities, including plants (Thomas and Parks 2025, 93–4). Ascribing the ability to communicate is a major hurdle to admitting plant personhood. It seems more likely that if you think you can communicate with someone or something that they have greater (or a different type of) moral value. In *Making Kin with Trees*, Solvejg Nitzke discusses both Robin Wall Kimmerer's ascription of personhood to plants as kin and Jared Farmer's exploration of old trees as persons and cared-fors (Nitzke 2025, 211, 228). Of course, Kimmerer's *Braiding Sweetgrass* (2013) is a major foundational text in the wave of popular critical plant studies and plant humanities texts. She weaves her cultural expertise as a Potawatomi Nation citizen and scientific expertise as an environmental biologist at SUNY to share her experiences with plants. There are even works uncovering plant personhood in the Western context (Scott 2023). In these cases, plant personhood is engaged as a serious theory.

Theoretical and Practical Concerns over Plant Personhood

As mentioned above, I worry that some will be unable to concede that the capabilities listed above are different in degree rather than kind when looking at the qualities of personhood characteristics in humans and plants. To this end, the first concern a skeptic might have is the science supporting Hall's claims about personhood capabilities. Trewavas is not without critics. Some of them are mainstream plant biologists who characterize the work of scientists like Trewavas as misleading and inaccurate. As Daniel Chamovitz summarizes:

> Controversy arose among plant biologists when a group of scientists who study various aspects of plant function defined a new field in 2005 that they called "plant neurobiology," which aims to study the information networks present in plants. These scientists saw many similarities between plant anatomy and physiology and the neural networks found in animals. (Chamovitz 2012, 136).

Chamovitz and others consider these similarities largely unfounded.

Within the ethics of care discussion earlier, personhood is reserved for the peak entities in a hierarchal value system and has a history of being coupled with faculties like reason and autonomy. A person is not to be trifled with, or used *merely* as a means to an end because they have inherent moral value based on their ability to reason. This ability in turn entitles them to dignity; traditionally, human dignity (Kant et al. 2019, G 4:429). And for many within the Anglo-European philosophical and scientific community, extending dignity (something traditionally only granted to human beings) to plants is nonsensical. As Chamovitz maintains:

> Without this caveat, anthropomorphism of plant behavior left unchecked can lead to unfortunate, if not humorous, consequences. For example, in 2008 the Swiss government established an ethics committee to protect the "dignity" of plants. A brainless plant likely does not worry about its dignity.
> (Chamovitz 2012, 140)

For many in the Western philosophical context, dignity's relationship to reason is very important; without the cognitive capacities associated with a human brain, or a similar enough brain to a human (those of some other animals), dignity does not make sense in the brain-less entity. Dignity, directly or indirectly, protects and honors brain-based cognition.

While Chamowitz finds using dignity in the context of plant life ridiculous on its face, Sabine Odparlik (2019) unpacks the promise of the dignity discourse for plants. Odparlik provides detailed nuance in contextualizing the purpose and power of the German term *Würde*, translated as dignity (Odparlik 2019, 60). While the Anglophone tradition follows Kantian use of dignity as only properly appropriate when discussing humans, the choice of dignity (*Würde*) was not a hapless or hurried decision in the Swiss law and Germanic discussions over the dignity of plant life (62–3). And despite differences of interpretation of the law, "(Supporters of the Dignity of Plants) all converge on at least the following point: plants have to be morally considered for their own sake, rather than because we use them and they please us" (62–3). And the ground for this mattering is not, as Chamovitz suggests in his popular text *What a Plant Knows*, founded on a shared ability to feel pleasure or pain (Chamovitz 2012, 139). Rather, it is a complex interrelation of autonomy, integrity, and having a good of one's own (Odparlik 2019, 63–5). There is not universal agreement in Germanic-speaking-traditions, by any means, on the dignity of plants. However, Odparlik's work shows that the philosophical debate over applying dignity to plant life is not purely perniciously anthropomorphic nor laughable.

So, claiming personhood for plant life is a radical departure from the Western philosophical tradition, as Hall shows. But it is unclear whether Hall's repeated assertions of plant intelligence and autonomy will satisfy entrenched of critics (e.g., Chamovitz) and those who think personhood is a uniquely human status. If one concedes to Hall and agrees to take stock in plant neurobiologist account of plant capabilities, they could still be deeply concerned about how one can practically live in the world if plants are persons. Hall's response is a reasonable one: this just points to problems in the way we see personhood morally as a hierarchy of different types of beings (Hall 2011, 157). Rather than argue that we must all be fruitarians, or creatures that only consume fruit and nuts, refraining from killing plants to survive, Hall argues that personhood cannot exempt use (160–1). Killing and consuming something so that one can live might seem like the epitome of using something merely as a means to an end—the end being your survival. Hall claims that we can still care for and share kinship with plant life, even if we must also destroy and consume plants. Rather than making such action explicitly immoral, he suggests a line similar to indigenous doctrine; that we minimize the destruction of plant life to what is needed, and that we promote plant life when at all possible.

Undoubtedly, critics of his response will have concerns, and they are concerns that go to the core of the moral considerability and plant life debate. If something has moral status or moral considerability, it usually means that something counts, morally speaking. Some things can have full or partial moral status and the moral concerns or rights of the entity with full moral status usually override the entity with partial or no moral status (Warren 1973, 57–8). Traditionally seen as entities with no moral status, or only instrumental moral value (i.e., only morally valuable in so far as they serve to promote or support something with full moral status or inherent moral value), plants are valuable because of their usefulness to humans (i.e., entities with full moral status or moral considerability.) But Hall resists this and even insists that a continuance of such instrumental thinking about plants is threatening to human existence.

> Purely instrumental relationships with plants are found to be ecologically destructive. The backgrounding of plants is dangerous because it severs opportunities for dialogical interaction between humans and the environments in which they live. Lacking in meaningful relationships of kinship, care, and solidarity, we risk complete human ecological dislocation. (Hall 2011, 14)

So we rely on valuing plants.

A great deal rests on the type of moral value an entity has. If we value things only in their use for humans, Hall claims that our "dislocation" from the intricate web of our ecological situatedness will be our undoing; not only in a physical sense, but also in a philosophical sense. Our very human identity is at stake.

> By distancing ourselves from plants and denying their autonomy, we jeopardize a true sense of human identity, situatedness, and responsibility. Only in the company of others do we arrive at the true sense of our own

> personhood and ecological identity. The risk we run by ignoring the personhood of plants is losing sight of the knowledge that we humans are dependent ecological beings.
>
> (Hall 2011, 14)

Perhaps like dehumanization in racism and sexism, (e.g., in denying the humanity in others, we are dehumanizing ourselves), we are undermining our own full personhood by denying personhood to deserving and qualifying entities. His alternative to this awareness disparity is moral personhood for plants.

Putting aside the highly contested field of "plant neurobiology," even the term and concept "plant behavior" is controversial in traditional plant biology.[1] If admitting that plants have behavior is controversial, how much more would be claiming that plants have the ability to sense and experience things, or have robust cognitive capacities that are on a spectrum with humans and other animals. That's simply not how dominant scientific or philosophical paradigms consider plants. They are so other to humans that it strains the credulity of a skeptic that plants would share capabilities with humans that are merely different in degree than in kind. These concerns may change in the future, but for the time being, another viable option for including plants as cared-fors might be a non-reason centered account of inherent moral value: Regan's subject-of-a-life.

Solution B: Plant Sentience and Subjects-of-a-Life

One important example of an ethicist that challenges personhood requirements to be more inclusive is Tom Regan. His "subject-of-a-life" criterion opens up the rationality-focused accounts of inherent moral value associated with Kant and other anthropocentric philosophies to include other animals. Instead of intelligence or autonomy being defining characteristics for inherent moral value, Regan associates the following with inherent moral value:

> beliefs and desires; perception, memory, and a sense of the future, including their own future; an emotional life together with the feelings of pleasure and pain; preference- and welfare-interests; the ability to initiate action in pursuit of their desires and goals, a psychophysical identity over time; and an individual welfare in the sense that their experiential life fares well or ill for them.
>
> (Regan 1983, 243)

For Regan, "subjects-of-a-life" are considered inherently morally valuable. They are entities with full moral status. Regan emerges from Kantian theory with a robust view that other animals belong in the coveted position of "ends in and of themselves" just as much as we do.

Importantly, subjects-of-a-life still have the core requirement of consciousness. There is a concern that if someone is not there (if there is not that "inner light" experiencing the world) then how can what happens to something matter: because it matters in virtue of it mattering to or for that thing. If how the world goes matters to you, then you have inherent moral value (Taylor 2011, 61–2). Kittay's daughter Sesha mentioned in Chapter 1 gives an example of this intuitive self from within (Kittay 2019, 250). That consciousness, the awareness of her experience that Sesha demonstrates with those who interact with her, illuminates her as a subject-of-a-life; what happens to her matters to her.

Regan's subject-of-a-life criterion is useful as a requirement for cared-fors in at least two important ways: it removes anthropocentric language and reason for inherent moral value, and it provides a convincing account of a moral patient who cannot be a moral agent. For plant cared-fors, these changes may appear to make care ethics much more accessible to more-than-humans.

Instead of citing Trewavas in this section of connecting subject-of-a-life criterion with the work of plant scientists, to diversify my evidence base I reference Paco Calvo, Natalie Lawrence, Stefano Mancuso, and Monica Gagliano. Paco Calvo is a philosopher of science and founder and director of the Minimal Intelligence Lab at the University of Murcia in Spain. Natalie Lawrence is Calvo's co-author of *Planta Sapiens* and has a background in philosophy of science. Stefano Mancuso is a botanist and prolific plant neurobiologist at the University of Florence. Monica Gagliano is an evolutionary ecologist at Southern Cross University's Biological Intelligence Lab. Not unlike Regan's original analysis, there is a large range of debate on whether these capacities exist in plants. Below, I address what each of these capacities might look like in plant life. Where appropriate, I provide examples from the scientific literature noted above.

Beliefs and Desires

This may be a difficult characteristic to show even in other animals, let alone in plants. For plants, some of the examples below might best show the capacity to have beliefs and desires. Beliefs would likely be present in cases of plants anticipating a future state of affairs. Desires might be present when plants pursue certain nutrients in their complex communities. Plant signaling in mycorrhizal networks suggests an understanding in the plants that chemical signaling will be understood and delivered throughout these networks in a way that will support the desires of the plant.

Perception

Plants perceive changes in their environment, as evinced by their responses to these changes in light, heat, moisture, soil composition, pest interaction and the signaling of other plants. Is this perception? It depends on your definition. If perception simply means the processing of the information provided by one's sensory organs, then plants can uncontroversially perceive things. If that requires the type of awareness and experience provided by a brain, maybe not.

Memory

Monica Gagliano (2018, 59) conducted a study on *Mimosa pudica* (also known as the shame plant) that suggests some plants may be able to show signs of adaptive learning. *Mimosa pudica* is a popular plant to use with plant testing because it is considered a sensitive plant—its leaves quickly retract when touched. In her experiment, Gagliano dropped a *Mimosa pudica* from a certain height (in a controlled drop using a pole). Initially, the plants appeared stunned, shutting their leaves quickly. But after several experiences, the plants became acclimated to the experience and realized that the dropping did not result in any harm. So, the retraction of leaves ceased in acculturated plants. This suggests that the plants were able to remember the experience, learn that the novel experience of falling did not endanger them, and respond accordingly in future cases.

Sense of the Future

Calvo and Lawrence identify a few cases that could indicate that plants have a sense of the future. For example, some tropical trees grow an excess of leaves before heavy rains come (Calvo and Lawrence 2023, 73). Plants that produce flowers prompt them to emerge at particular times that predict when pollinators will be present (77). Preparations made for future seasons, years, and generations might also suggest this awareness of a future.

Emotional Life with Feelings of Pleasure and Pain

This requirement is a potential challenge for plants, as ascribing emotions to them sounds incredibly controversial. Not unlike intelligence being associated with a brain, emotions likely also require a complex cognitive system. In addressing this, Calvo and Lawrence identify suffering and pain as generally evolutionarily useful (Calvo and Lawrence 2023, 188). Some of the chemical compounds that appear in plants function to alleviate pain and moderate behavior when their analogues appear in humans and other animals (e.g., melatonin).

> Some of these chemicals are only produced in situations when plants are stressed or injured. Plants make many substances that have painkilling or anaesthetic, such as ethylene. Ethylene seems to be an important stress signal in bacteria, fungi and lichens as well: its message bridges large swathes of the evolutionary tree.
>
> (Calvo and Lawrence 2023, 188)

In this approach, Calvo and Lawrence look for chemicals released in contexts that might suggest a particular function. Given that we humans are not able to access the experience of other beings, perhaps this is a wise way of approaching the problem. Calvo and Lawrence's functional approach may convince some of the possibility of suffering in brain-less, biotic creatures.

Preference and Welfare Interests

As discussed in the personhood section, a plant's ability to choose one option that corresponds with its well-being would suggest that preferences and welfare interests exist for plants. Examples of this appear in the personhood autonomy discussion above, where a plant chooses a soil option that is better for it (e.g., a less occupied space) than a less optimal option (e.g., a crowded landscape). Acting on a growth option that is better for the plant's living (e.g., producing less stress than another option) demonstrates both the existence of preferences and their ability to correspond with the plant's welfare interests.

Ability to Initiate Actions in Pursuit of Desires and Goals

One important example of this requirement is the Cornish mallow. Calvo and Lawrence understand the shift of Cornish mallow leaf orientation overnight to a position that can quickly get strong light from the morning sun as a way of anticipating the light, not merely responding to stimulus. This ability to anticipate for the attainment of basic desires (necessary and appropriate amounts of nutrients, a supportive non-hostile environment, etc.) which also serve larger goals (living, growing, expanding, reproducing, etc.) demonstrates this requirement.

Psychophysical Identity over Time

Plant memory and the ability to differentiate self and other suggests a basic understanding of identity over time. Root awareness of self-roots as opposed to non-self- roots might be enough to meet this criterion (Trewavas 2014, 188). The psycho- portion of psychophysical identity might be more controversial if psychological identity requires the presence of a brain. But the ability of an organism to differentiate between self and other is an indicator of some awareness of embodiment over time that could satisfy this condition.

Individual Welfare Related to their Experience

I think that this capacity corresponds with the pain and suffering one. The plant needs to be able to experience its life (i.e., be conscious) and have that experience relate to the plant's quality of life in a morally relevant way. This relies on consciousness and the subject-of-a-life's ability to have a good life in light of its pain and pleasure experiences. This is the crux of the problem facing theorists who want to include plants as subjects-of-a-life; to be a subject-of-a-life, the organism must be robustly conscious.

Concerns over Plant Sentience and Subjects-of-a-Life

The emphasis of the subject-of-a-life criteria is the subject, as in, the locus of perspectival awareness or consciousness. The first step in approaching Regan's subject-of-a-life criterion, granting consciousness to plants, may be an impasse for skeptics. Despite consciousness still being a puzzle itself, it has played a very important role in deciding who or what has moral status. When considering whether plants are conscious, there are two particular points of concern: the brain and subjectivity.

Brains and Plant Neurobiology

As mentioned before, the area of plant consciousness is very controversial. Even popular books like *What a Plant Thinks*, rich with anthropomorphic language, end with a disclaimer that all of the "senses" that the book discusses, and details are not even close to what humans and other sentient beings do.

> If we keep in mind that a plant doesn't have a brain, it follows then that any anthropomorphic description is at its base severely limited. It allows us to continue to anthropomorphize behavior for the sake of literary clarity while remembering that all such descriptions must be tempered by the idea of a brainless plant. While we used the same terms – "see," "smell," "feel" – we also know that the overall sensual experience is qualitatively different for plants and people.
>
> (Chamovitz 2012, 139–40)

Sentience means awareness or being conscious of something. Often in the animal ethics literature, it simply means pain/pleasure awareness, or the ability to suffer. But it could also be related to awareness through different senses.

Plant neurobiologists like Paco Calvo focus on function when considering plant consciousness. Calvo does not argue that plants have neurons or brains, but rather vegetal cells and systems that provide similar functions. But similar functions may not be enough to convince skeptical fellow plant physiologists like Chamovitz. Those scholars need brains, and plants do not have them.

Subjectivity and Plants

Aside from the issue of the brain and its associated capacities, subjectivity and the appearance of a single subject of experience is another issue when considering skepticism about plant consciousness. Regan's subject-of-a-life occurs within the animal ethics literature as a way of including clearly conscious animals in an otherwise anthropocentric ethical theory. The capacities required to qualify as subjects-of-a-life may be considered zoocentric, rather than anthropocentric, because consciousness is so central to how we understand animals and correlated moral value.

If you can suffer, how the world goes for you likely matters to you (Singer 1989, 154). But what if there is no reflexive self-awareness, no inner *sense* of you? How can humans conceptualize a cared-for that may not have a centralized subjectivity? Plants exist as such wildly different beings from us. If we grant that there is some minimal sentience occurring, it likely does not process in a central location like in humans and some other animals. *You* via perspectival awareness have a visceral stake in avoiding pain and seeking pleasure. However, if there's no you, one might wonder if you can't suffer, if there's nothing-it's-like to be you, does it matter how the world goes for you? (Sandler 2019, 41) The sentiocentrist would say no. So, in losing this subject-of-experience, the skeptic would argue we lose the part of the being that makes an organism matter in and of itself morally speaking.

Because there is no brain or no central point of perspectival awareness, plant identity and self-hood may be too different from ours to easily map onto our understanding of subjectivity. There's an aesthetic salience to animal subjectivity, a focus on a particular perspective and a singular being. Plant consciousness, if it exists, is likely diffused in a completely different way because of the absence of a central processing hub (Ponteau 2019, 86). It is easier for humans to understand plant identity from the outside, observing where one body begins and another ends. This difference in way-of-being need not be an impediment, but might be too different and interesting to be reduced to an animal-centered understanding of subjectivity, perspectival awareness focused on one point.

Belief in plant consciousness depends in large part on one's paradigm. If you are starting from a neo-pagan paradigm, plant sentience (and maybe even personhood) might appear intuitive and attractive. If you are starting from an American Protestant Christian view that emphasizes humanity's right to have dominion over all of creation, perhaps not. But given these starkly different paradigms that human individuals within the web of care are working from, to make the case for the domain of care extending to plants, additional options for grounding intrinsic moral value are useful. This leads to a possible third response to the anthropocentric requirements for cared-fors: avoid personhood and subjects-of-a-life and consider an alternative.

Solution C: Plants as Living Beings

If plants are not persons or conscious, how can they be cared-fors in care ethics? At this point, one should return to the question posed in Chapter 1: what is anthropocentrism's function in care ethics, and can a collection of other characteristics stand-in for anthropocentrist requirements for the cared-for? I think one of the primary functions of personhood and humanity requirements in the cared-for is the establishment of inherent moral value. Whether appealing to personhood or human-centric, in contemporary contexts I think the majority view is that if anything has moral value in and of itself, it would be persons and humans. So, by cordoning off all ethically relevant cared-fors as persons and humans, an assumption of inherent moral value is built into ethically caring relationship. Aside from this, capabilities associated with personhood and subjects-of-a-life ensure that the participants in the caring relationships are able to engage in the relationship in such a way that ethical care is possible. As noted in Chapter 1, the caring relationship requires actions on both the part of the carer and the cared-for. If both are persons, you can be assured that they have the requisite capacities for care, reciprocity, and so forth. Subjects-of-a-life can make the same assurances. But if the cared-for is not a person or a subject-of-a-life, how can one make sure that care is possible? What could possibly qualify a non-sentient, non-person plant as a moral patient? In this section, I offer biocentrism as a way that non-sentient non-person could have inherent moral value that would qualify it as a cared-for. The most promising sources of inherent moral value for the plant cared-for could be autonomy, narrative identity, and biocentrism.

Contrary to Nodding's intuition about the hermit monk out in nature being devoid of ethically caring relationships, many environmental philosophers would grant that a lone human surrounded by various forms of non-human life could have rich moral relationships and obligations in such a seemingly isolated lifestyle in a wild place (Noddings 2013, 80). My argument in favor of a plant inclusive ethics of care may rest on showing that the non-sentient, non-person other in a caring relationship has inherent moral value and can have care-relevant capacities that can make sense of the structure of care ethics, without having the benefit of importing the moral characteristics and qualities of a sentient animal, human, or person. Biocentrism is one possible way to address this concern. For biocentrist accounts, all living beings have intrinsic moral value. There are different ways that each theorist seeks to do this.

One way to establish inherent moral value is by demonstrating that plants have interests. Ronald Sandler offers an account of how plants can have interests while noting that having interests is usually associated with conscious beings, but he provides a way that plants can have interests without being conscious (Sandler 2019, 41). For plants, these are biological interests, rather than the psychological interests most commonly associated with inherent moral value.

> The most common explanation for what grounds the interests of non-conscious organisms (as well as the bare biological interests of conscious ones) is that the organisms are goal-directed systems (Goodpaster 1978; Sandler 2007; Taylor 1986; Varner 1998). Their parts and processes are organized in ways and for reasons related to accomplishing things such as resisting predators and reproducing.
>
> (Sandler 2019, 42)

And so, "Flowers produce nectar in order to attract pollinators; ivy has tendrils in order that it can climb; tree roots are skototropic in order that they grow deeper into the ground" (Sandler 2019, 42). Again, we find that language is key for understanding this claim and avoiding pernicious anthropomorphism. Put another way, "That nonconscious organisms are end or goal-directed systems means that it is possible to affect them in ways that are conducive or detrimental to accomplishing their goals or ends—that is to harm and benefit them" (42). Sandler explains that this goal-direction and end-orientation could establish the living organism as morally valuable in and of itself, regardless of its usefulness to humans or other animals, if one resists the wall of moral difference erected between sentient and non-sentient organisms (43–4). By pushing against the value difference between psychological interests and biological interests, he argues that one could show both types of interests should count when considering well-being and moral value.

One example of a rich end-focused biocentric account is Paul W. Taylor's argument for grounding respect for biotic individuals in telos (Taylor 2011, 44–5). Respect, an attitude often used to respond to an entity with dignity, requires an acknowledgement of the telos of the other: that the biotic individual has a goal, purpose, and way of flourishing that stands distinct from a human being and that has value in and of itself. For Taylor, having a telos grants a biotic individual moral considerability. And importantly for Taylor, only natural biotic organisms have telos. Domesticated living things have had their natural telos perverted by human goals and interference and are now artificial (5–6). As we'll see in Chapter 6, the same could easily be said of domestic tomato plants. This bifurcation between "natural" living things and those who have been "manipulated" by humans such that their telos has been re-aligned or disregarded entirely to fit an artificial end determined by humans.

Telos is highlighting the aims of a biological individual and connecting those aims to some internal purpose, function, or meaning, inherent worth for Taylor (Taylor 2011, 75–6). Yet, for Taylor, having a telos elicits the same moral attitude, respect, that others argue should be elicited by things like human dignity and personhood. One might return to the work that anthropocentric requirements are doing in care ethics; identifying who or what has inherent moral value and which capabilities require special treatment. For Taylor (2011), Katz (2002), and the like, it requires us not to impede upon such things, as human interference often undermines a wild living plant's ability to flourish. But some accounts of telos's inability to account for human aims existing alongside wild organisms' aims make it difficult to accept in a world like ours. The wild and domestic distinction is not so easily drawn. Leaving behind domestic creatures as not morally relevant is another major drawback of this approach that may not follow one's intuition. But that could be remedied by arguing for a telos grounded in living rather than absence of human intervention.

Some might find telos attractive in part because these nonconscious beings are living and have a long, evolutionary history. The biocentrist etiological view proposes that this is how we get a morally relevant concept of biological function in light of our contemporary scientific paradigms that roughly reject intelligent design or creation stories that make sense of function talk. For example, it makes more sense to think of natural processes as functions if there is some design in place for the organism's ultimate goal in life. These biological processes support the realization of that end goal. However, without the cosmology of some design, we are left with natural processes and a history of using function talk. This is why understanding function is so crucial to biocentrists. Where do these goals come from? Why do we consider them goals? One way to address this is the etiological view which proposes roughly that we can speak meaningfully of biological functions by looking backward at evolutionary processes. Through natural selection processes, we can make sense of functions (in light of what historically produced the current living organism) that connect with the organism's interests (Sandler 2019, 42).

In recent literature, there are major concerns about biocentrism and the move from description to prescription involved with function to interest talk. In one such example, John Basl argues that biocentrism is dead, in light of alternative non-life-based systems that could also have a welfare (Basl 2019). The arguments for living organisms having a goal or end also seem to apply to some non-living systems, and so why are living ones alone valued while non-living ones are discarded? Katie McShane argues that the etiological functions of an organism are not strongly related to an organism having interests in part because those activities that enabled the success of the organism historically may not be benefiting a particular organism now (McShane 2019, 3507). As McShane concludes, the "prospects (of an account) depend(s) on whether they can justify the conceptual relationships that they posit between possessing certain characteristics and having a welfare" (3514). Sentience has been considered such a characteristic that links having a welfare, and something being harmful or beneficial to an organism.

Another important criticism of biocentrism and the self-sustaining, or autopoietic, account of interests is given by Stephanie Hoffmann (2022, 229). She argues that the identity of the organism is key to determining whether interest speak is even relevant because living beings have such disparate ways of being individuated. It is easier to understand sentient beings as having interests in part because the "self" that they have is a point of perspectival consciousness that connects to the bits and pieces that comprise a whole organism. Without sentience, as mentioned above, it is unclear one can have selfhood. Without selfhood, Hoffmann shows that individuation is the starting point of determining interests. But through her process of considering different biocentric options for individuation, each seems to let in too many things that could have interests or too many possible interests (229). And so, with no account of a biological individual, there seems like there can be no meaningful way forward to biocentric accounts that purport to grant inherent moral value to biotic individuals (229). Despite these concerns, some consider biocentrism to still be a viable candidate for explaining both the source of intrinsic moral value and capabilities necessary for care.

Solution D: Plants as Flourishing Beings

Were one to reject biocentrism, are there any characteristics that can satisfyingly connect a living organism and its welfare? Angela Kallhoff offers such an account on what she considers an alternative to biocentric arguments by focusing first and foremost on the otherness of plant life. She posits that their flourishing and ability to be harmed help make sense of a plant having a good life. This is especially important for the context of this project as Kittay requires the cared-for to be able to flourish in her care ethics. While Kittay restricts the ability to flourish to human persons, Kallhoff provides an account of plant flourishing (Kallhoff 2019, 53). According to Kallhoff, "a plant flourishes when it realizes a full life-cycle, when it develops its characteristics during this process, and when its vitality (its continued stress-responsiveness without loss of capacities to react strategically and without illness) is kept intact" (53). This approach to establishing flourishing is an attempt to avoid anthropomorphism and ground flourishing in plant-relevant aspects of plant life. From this starting point, actions taken towards the plant can be understood as being beneficial or harmful to them in light of their ability to exercise certain capabilities that comprise the good life for that plant (54).

As Kallhoff explains,

> (when connected to persons) flourishing is explored as a basic concept of the good life and one in light with concepts such as 'happiness', 'well-being' and the like. To say this concept is useful in addressing plants does not say that the set of underlying natural capacities is the same in persons and plants. Instead, it says that in order to explain the good life of those living beings which unfold and tend to realize a specific life-form, it is necessary to look at the biological side as well.
>
> (Kallhoff 2014, 689)

Chris Cuomo is another ecological theorist who supports flourishing as a more basic foundation for environmental ethics. Though Cuomo includes whole systems in her account of ecological feminist ethics as an ethic of flourishing which, like Kallhoff emerges from Aristotelian ethics (Cuomo 1998).

While this approach may not be entirely satisfying to the critiques aimed at biocentrism, it takes seriously the tension between biological and moral causation and relation in a way that fits with the larger normative framework of care ethicists like Kittay. I think skeptics might be more ready to admit that plants can flourish. But critics of Aristotelian telos, or goals, may have difficulty accepting this claim. Additionally, those who consider flourishing to be an activity or state that is only accessible to human beings may deny this as a possibility for plant life. Others might worry that depending on how flourishing is defined, one might slip down a slope that affirms more than plants can flourish—including non-living things and systems. While this might not be of particular concern for my immediate project here, it may be enough of a concern for others that flourishing is not a viable candidate for establishing intrinsic moral value and capabilities relevant to care.

Solution E: Plants as Kin

Arguably related in some way to the early parts of this chapter (e.g., Hall's contemporary account of personhood emerging from Darwinian scientific theory), kinship accounts of plant moral value rest on both our shared lineage and our relationships. This differentiates these views from biocentrism and Kalhoff's flourishing because the important factor in grounding moral consideration is relatedness. In being kin, there is a connection between two things, no matter how disparate. That connection is not trivial; it sets apart the related entities as valuable in a way that non-kin are not, but also relatedness entails responsibilities and obligations (Williams 2018, 126–7). Many kinship accounts come from indigenous systems of thought (Salmón 2020, Tynan 2021). Though they are also gaining ground in non-indigenous communities (e.g., Donna Haraway's *Staying with the Trouble: Making Kin in the Cthulucene*).

One example of kinship from indigenous thought is found in *Kayanerenkó:wa The Great Law of Peace* by Kayanesenh Paul Williams. The Great Law is the foundational creation of the Iroquois Confederation. It provides the laws, traditions, and cultural practices laid out by the Peacemaker who united the Seneca, Cayuga, Onondaga, Oneida, and Mohawk tribes after a period of great conflict and civil war. The Great Law tells the story of the Peacemaker and lays out inter-tribal relations for a peaceful coexistence. Within these traditions, kinship with natural entities, land, and Earth is essential. Given that humans are in relationship with the feminine earth, sometimes affectionately called Mother Earth, there are responsibilities that we have in response to that relationship. As kin, the natural environment is not separate from, but identified with the human. Plants, animals, and natural entities like rivers can also be in kinship relations with individual and groups of humans.

This is one of the most profoundly different ways to ground moral value amongst the options provided. Rather than focus on the individual capabilities of the cared-for, kinship centers relatedness as the determining factor of responsibility. In this radical way, it is in keeping with the central shift in care ethics—a move from centering the individual, to centering the relationship. Also, to be in relationship, capabilities may be assumed. Often, in indigenous contexts, the natural entities that humans are in relationship with are parts of an ecological community. However, kinship approaches are not yet mainstream in care ethics. There may also be worldview restrictions on who could affirm kinship with plants. For example, if someone does not accept a gendered version of the earth as provider or does not believe in creation myths that explain the relation of humans to other beings, it might be difficult if not untenable for some worldviews to espouse kinship.

There may be technical issues around how exactly kinship works. What grounds kinship? Can kinship include relationships with anything, including machines or artificial creations? What is the process of becoming kin? Is it available for anyone, or is it only coherent within a particular cultural or religious context? These are important questions and concerns. But if kinship were accepted, it could explain how plants could have intrinsic moral value and care relevant capabilities, as the ability to be in relation is a key feature of care ethics.

Conclusion: A Buffet of Plant Cared-for Options

I think all of these approaches for grounding the moral value and relevant capabilities to plant cared-fors have issues and none is perfectly unassailable by critics. But that is the nature of practical ethics. I provide this wide range of possibilities in part to demonstrate the wide breadth of resources an ethicist has at their disposal if they would like to include plants in a care ethics framework. I also want to allow space for varied practitioners to access care ethics and be able to apply it in their work and lives. Practical application and connection with the work of practitioners is very important to me, as a theorist. I think it's all well and good to have an elegant theory and I also want to provide space for value pluralism and pragmatism. I agree with McShane (2007) that one's account of intrinsic moral value may have radical implications for the contours of their ethical theory. But I also do not feel comfortable mandating that all must share my view. Each approach listed above can at least grant the two essential factors for making sure that the cared-for makes sense within an ethically caring relationship: intrinsic moral value and care-relevant capabilities. Interestingly, the characteristics or qualities that qualify something as having intrinsic moral value tend to also be care-relevant capabilities. For example, in personhood accounts, the ability to act and have autonomy are capabilities that support a caring relationship. Being able to act might make sense of the plant cared-for's ability to respond to the care offered by the carer. In having autonomy, a plant can either accept or reject the actions of the carer. Similarly, the ability to feel and communicate, required by Regan's subject-of-a-life make claims of the possibility of a caring relationship between plants and humans reasonable. It also allows for emotion being present in the caring relationship not only from the carer but also the cared-for. For biocentrist accounts, living things respond to their environments. This ability to respond could qualify the plant as being capable of responsive interactivity in the caring relationship, as well as being able to demonstrate the success or failure of care given by the carer. Flourishing is a necessary part of Kittay's account of care ethics, and the ability of a plant to flourish could fulfill the success condition requirement of care in her ethical theory. Finally, the bottom floor of kinship accounts is relationship. Were plants to be considered kin, they would be able to participate in relation to human beings, not only explaining their intrinsic moral value, but also providing the key component of care ethics: relationship.

I think the view one asserts reveals more about that person and their beliefs about the qualities and characteristics of plants than the nature of the plants themselves. The plant humanities and critical plant studies community provides a wide range of belief sets about what plants are and what they can do. In keeping with that plurality of beliefs, I wanted to provide a map of approaches that could make sense of care ethics under a range of paradigms. I think the precautionary principle could be useful here, as well. For those readers who find consciousness too controversial at this point, one benefit of taking options (c), (d), or (e) is that if public and scientific understanding of plant capabilities were to shift (e.g., were plants to be found conscious), under those approaches that do not require consciousness, the plant would already be considered a cared-for.

Another benefit to taking the routes of (c), (d), or (e) is that it establishes that plants have inherent moral value without disallowing any potential of distinguishing hierarchies or differences of treatment on the identity of the cared for. So, the baseline of action is to avoid harming plants. There may be good reasons that one might need to end the life of a plant, which the normative ethical theory (in my case, care ethics) can sort through and determine. But the default position is to avoid harming plants because they have intrinsic moral value. This is an intuition captured in Richard Sylvan's The Last Man thought experiment where one imagines that there is one human left on earth (Routley 1973). With that human's last days and moments, she could destroy everything else on Earth—but no humans will be harmed or affected because she's the last one. If only humans have inherent moral value, then is there anything wrong in this last act of destruction? Biocentrists identify the intuition being pressed that there's something wrong about the decision to destroy it because living things remain and have inherent moral value (Routley 1973). I admit that I am very swayed by this thought experiment. The intuition that there is still something inherently valuable in a world where humans and sentient animals are no longer present resonates with me. The particular argument finding the source of that value without sentience may be elusive, but I can't help but be convinced that there is something of inherent moral value in such a world.

In the next chapter, I will provide a more complete account of how a plant cared-for can be coherent with an ethic of care. I will demonstrate that the internal function anthropocentrism has played in care ethics theory can be fulfilled by non-human plant cared-fors and respond to objections.

Note

1 For example, the academic journal *Plant Signaling and Behavior* published by Taylor and Francis began in 2006 as a project of the Society of Plant Neurobiology. A 2009 vote changed the title of the society and journal to *Plant Signaling and Behavior*, a less controversial and more acceptable way of framing the work in the journal. As the information on the publication's website states: "The goal of this field is to illuminate the structure of the information network that exists within

plants. Plants are dynamic and highly sensitive organisms that actively and competitively forage for limited resources both above and below ground. Plants accurately compute inputs from the environment, use sophisticated cost-benefit analysis, and take action to mitigate diverse environmental insults. Plants are also capable of refined recognition of self and non-self, and are territorial in behavior. This view sees plants as information processing organisms with complex, long-distance communication systems within the plant body and extending into the surrounding ecosystem. Our Society was originally founded in 2005 as the Society for Plant Neurobiology to reflect these views of plant function. In May 2009 the Society voted to expand its view and change its name accordingly.

One goal of establishing a community for Plant Signaling and Behavior is to provide a venue for all interested biologists to explore complex plant behavior utilizing all levels of experimental approach. Among our symposia participants have been molecular geneticists, biochemists, electrophysiologists, physiological ecologists, community ecologists, mathematical modelers, plant designers, and even philosophers. Plant Signaling and Behavior will use the lens of integrated signaling, communication, and behavior to integrate data obtained at the genetic, molecular, biochemical and cellular levels with physiology, development and behavior of individual organisms, plant ecosystems and evolution" (www-tandfonline-com.libproxy.temple.edu/action/journalInformation?show=societyInformation&journalCode=kpsb20). Work associated with the journal averages in the second quadrant of research quality according to Scimago Journal and Country Rank (www.scimagojr.com/journalsearch.php?q=8200153106&tip=sid&clean=0). So, the work that adheres to the journal's mission is not as firmly central to contemporary scientific knowledge powerhouses like *Nature*, but it is not so fringe as to be in the fourth quadrant or unranked entirely.

References

Bader, M.K.-F. and S. Leuzinger. "Hydraulic Coupling of a Leafless Kauri Tree Remnant to Conspecific Hosts." *iScience*, 19 (September 2019), 1238–1243.

Baier, Annette. "A Naturalist View of Persons." *Proceedings and Addresses of the American Philosophical Association*, 65(3) (1991), 5–17.

Baluška, František, Stefano Mancuso, and Dieter Volkmann (eds). *Communication in Plants.* Springer, 2006.

Basl, John. *The Death of the Ethic of Life.* Oxford University Press, 2019.

Bekoff, Marc. "On Respecting Plants: A Conversation with Dr. Marc Beckoff." *The Plant Initiative Podcast* (December 22, 2022). https://youtu.be/pxpc-7HWgygN92

Calvo, Paco and Natalie Lawrence. *Planta Sapiens: The New Science of Plant Intelligence* W.W. Norton & Company, Inc., 2023.

Calvo, Paco, Vicente Raja, and Miguel Segundo-Ortin. "Don't Jump the Gun Quite Yet: Aiming for the True Target in Plant Neurobiology." *Protoplasma*, 262 (2025), 247–254.

Chamovitz, Daniel. *What a Plant Knows: A Field Guide to the Senses.* Scientific American, 2012.

Cuomo, Chris J. *Feminism and Ecological Communities: An Ethic of Flourishing.* Routledge, 1998.

Gagliano, Monica. *Thus Spoke the Plant: A Remarkable Journey of Groundbreaking Scientific Discoveries and Personal Encounters with Plants.* North Atlantic Books, 2018.

Goodpaster, Kenneth E. "On Being Morally Considerable." *The Journal of Philosophy*, 75(6) (1978), 308–325.

Gruen, Lori. *Ethics and Animals: An Introduction*. Cambridge University Press, 2011.

Hall, Matthew. *Plants as Persons: A Philosophical Botany*. State University of New York Press, 2011.

Haraway, Donna J. *Staying with the Trouble: Making Kin in the Chthulucene*. Duke University Press, 2016.

Hoffmann, Stephanie. "An Evaluation of the Autopoietic Account of Interests." *Synthese*, 200(3) (2022), 1–18.

Hutton, Christopher. *Integrationism and the Self: Reflections on the Legal Personhood of Animals*. Routledge, 2019.

Kallhoff, Angela. "Plants in Ethics: Why Flourishing Deserves Moral Respect." *Environmental Values*, 23(6) (2014), 685–700.

Kallhoff, Angela. "The Flourishing of Plants: A Neo-Aristotelian Approach to Plant Ethics." In Angela Kallhoff, Marcello Di Paola and Maria Schörgenhumer (eds), *Plant Ethics: Concepts and Applications* (pp. 51–58). Routledge, 2019.

Kant, I., C. Bennett, J. Saunders, and R. Stern. *Groundwork for the Metaphysics of Morals*. Oxford University Press, 2019.

Katz, Eric. "The Liberation of Humanity and Nature." *Environmental Values*, 11(4) (2002), 397–405.

Kimmerer, Robin Wall. *Braiding Sweetgrass: Indigenous Wisdom, Scientific Knowledge, and the Teachings of Plants*. Milkweed Editions, 2013.

Kingsland, Sharon E. and Lincoln Taiz. "Plant 'intelligence' and the misuse of historical sources as evidence." *Protoplasma*, 262 (2025): 223–246.

Kittay, Eva Feder. *Learning from My Daughter: Valuing Disabled Minds and Caring That Matters*. Oxford University Press, 2019.

Mancuso, Stefano, and Alessandra Viola. *Brilliant Green: The Surprising History and Science of Plant Intelligence*. Island Press, 2015.

McShane, Katie. "Why Environmental Ethics Shouldn't Give Up on Intrinsic Value." *Environmental Ethics*, 29(1) (2007), 43–61.

McShane, Katie. "Against Etiological Function Accounts of Interests." *Synthese*, 198 (4) (2019), 3499–3517.

Nitzke, Solvejg. *Making Kin with Trees: A Cultural Poetics of Interspecies Care*. Palgrave MacMillan, 2025.

Noddings, Nel. *Caring: A Relational Approach to Ethics & Moral Education*, 2nd edn. University of California Press, 2013.

Odparlik, Sabine. "The Dignity of Plants: An Overview of the Discussion in German-Speaking Countries." In Angela Kallhoff, Marcello Di Paola and Maria Schörgenhumer (eds), *Plant Ethics: Concepts and Applications* (pp. 59–69). Routledge, 2019.

Palmer, Clare. *Animal Ethics in Context: a relational approach*. Columbia University Press, 2010.

Pitt, Jon. *Botanical Imagination: Rethinking Plants in Modern Japan*. Cornell University Press, 2025.

Plumwood, Val. *Feminism and the Mastery of Nature*. Routledge, 2002.

Ponteau, Sylvie. "Plants as Open Beings: From Aesthetics to plant-human ethics." In Angela Kallhoff, Marcello Di Paola and Maria Schörgenhumer (eds), *Plant Ethics: Concepts and Applications* (pp. 82–97). Routledge, 2019.

The Royal Society. "Professor Anthony Trewavas, RFC." Fellows Directory. https://royalsociety.org/people/anthony-trewavas-12436/ (accessed November 2025).

Regan, Tom. *The Case for Animal Rights*. University of California Press, 1983.

Rolston, HolmesIII. *Environmental Ethics: Duties to and Values in the Natural World.* Temple University Press, 1988.

Routley, Richard. "Is There a Need For a New, an Environmental, Ethic?" *Proceedings of the XVth World Congress of Philosophy*, 1 (1973), 205–210.

Salmón, Enrique. *Iwígara: The Kinship of Plants and People.* Timber Press, 2020.

Sandler, Ronald. *Character and Environment: A Virtue-Oriented Approach to Environmental Ethics.* Columbia University Press, 2007.

Sandler, Ronald. "Is Considering the Interests of Plants Absurd?" In Angela Kallhoff, Marcello Di Paola and Maria Schörgenhumer (eds), *Plant Ethics: Concepts and Applications* (pp. 40–50). Routledge, 2019.

Science Museum Group. "Trewavas, Anthony J." https://collection.sciencemuseum group.org.uk/people/ap27253/trewavas-anthony-j.

Scott, Natalia Schwien. "The Plant Delighteth: Plant Personhood in the Study of Western Esotericism." *The Pomegranante*, 25(1) (2023), 167–185.

Singer, Peter. "All Animals are Equal." In Tom Regan and Peter Singer (eds), *Animal Rights and Human Obligations* (pp. 215–226). Cambridge University Press, 1989.

Taylor, Paul W. *Respect for Nature: A Theory of Environmental Ethics*, 25th Anniversary Edition. Princeton University Press, 2011.

Taylor, Paul W. "The Ethics of Respect for Nature." *Environmental Ethics*, 3(3) (1981), 197–218.

Thomas, Mariko O. and Melissa M. Parks. *Storying Plant Communication: More-Than-Human Relationships in New Mexico.* Bloomsbury Academic, 2025.

Tompkins, Peter and Christopher Bird. *The Secret Life of Plants.* Harper and Row, 1973.

Trewavas, Anthony. *Plant Behaviour and Intelligence.* Oxford University Press, 2014.

Trewavas, Anthony. "Plant Intelligence Dux: a comprehensive rebuttal of Kingsland and Taiz." *Protoplasma*, 262 (2025), 255–266.

Tynan, Lauren. "What is relationality? Indigenous knowledge, practices and responsibilities with kin." *Cultural Geographies*, 28(4) (2021), 597–610.

Varncr, Gary E. *Personhood, Ethics, and Animal Cognition: Situating Animals in Hare's Two Level Utilitarianism.* Oxford University Press, 2012.

Warren, Mary Anne. "On the Moral and Legal Status of Abortion." *The Monist, 57 (4)* (1973), 53–4.

Warren, Mary Anne. *Moral Status: Obligations to Persons and Other Living Things.* Clarendon Press, 1997.

Williams, Kayanesenh Paul. *Kayanerenkó:wa The Great Law of Peace.* University of Manitoba Press, 2018.

3 Defending a Plant-Inclusive Care Ethic

Ethical Care in Relationships with Plants

Can plant cared-fors fit within care ethics? So much of what Noddings, Kittay, and Held claim as core to ethical care may appear only accessible to human person cared-fors; Noddings' reciprocity, Kittay's relational identity and care-success indicators, and Held's framework can easily exist within human relationships with human cared-fors. So, is it even possible that a caring relationship with a plant could be coherent within ethics? In this chapter, I argue that it can.

After contending that there are plausible arguments for intrinsic moral value that ground establishing plants as cared-fors in Chapter 2, I turn to the coherency of plant cared-fors in care ethics. I identify what I consider to be core ethical structures of the relationship between carer and cared-for in care ethics that might be challenging: the attitude of the carer for the cared-for, a caring relationship that includes caring affect, dependency, and responsive interactivity, and success conditions that can benefit the cared-for. After discussing the fit of plant cared-fors within this structure, I address a number of concerns that an interlocutor might have in response to my account.

By opening up ethical care beyond the human community, we can acknowledge the moral weightiness of the relationships we humans have with more than human non-persons and transform our understanding of our place in the world, as dependent creatures capable of ethical care within our communities, which are comprised of persons and non-persons from the start.

Verifying that non-person plant cared-fors fit within care ethics requires examining them within core features of the normative theory. Keeping a focus on care as a practice centered around certain types of relationships, the caring relationships that I am concerned with in a plant-inclusive ethics of care require attention to the following:

1 Attitude of the carer that ascribes intrinsic moral value to the cared-for,
2 Presence of caring affect, social dependency, and responsive interactivity in the relationship between carer and cared-for, and
3 Possibility of beneficial consequences of caring action for the cared-for.

DOI: 10.4324/9781003603665-4

Ensuring that plant cared-fors are coherent with these three components of care ethics can make sense of non-person plant cared-fors within care ethics.

Valuing-Attitudes

Common attitudes that carers have for their cared-fors in caring relationships include love, care, consideration, attentiveness, and commitment (Noddings 2013, 19–20, 63). These attitudes can motivate the carer to engage in caring actions. Valuing-attitudes also orient the carer in a caring relationship and ensure that a cared-for is an appropriate fit for the ethical structure of care. I want to emphasize valuing-attitudes towards the cared-for for two primary reasons: establishing the orientation of carer to the cared-for and enabling pragmatic value pluralism.

First, attitudes from carers orient the moral agent towards the cared-for. It is useful to bring in Katie McShane's work on intrinsic moral value and some valuing-attitudes and actions. McShane argues that, in part, some attitudes that humans take towards non-human others require an assumption of moral value on the part of the agent (McShane 2007, 43–61). So, if I love my dog in a certain way, my attitude of love is evidence of me ascribing an intrinsic moral value to my dog (52). Without this ascription of moral value, it is impossible to capture exactly what is happening with that attitude of love. The self-sacrifice that follows from a valuing-attitude in this case, would be mistaken were my dog not to have intrinsic moral value (54). But McShane urges us to resist assuming right away it is a mistake and rather evaluate whether the attitude is correct. For example, the cultural phenomenon of plant parenthood is apparent on digital social networking outlets (Cabrera 2018). It would be easy to dismiss these instances of media users posting about loving their plant kids as a clear mistaken ascription of value and an instance of anthropomorphism. However, if the plant "child" does indeed have inherent moral value (as discussed in Chapter 2), then perhaps this attitude is correctly orienting the human "parent" towards her plant cared-for. The coherency of a valuing-attitude correctly picking out the moral value in a plant cared-for and fitting within the ethical structure of care ethics is clear. Humans can surely have love for individual plants in the same way that they do for other humans and other animals.

While love can help pick out an intrinsically valuable cared-for, the correctness of this judgment can be supported by an account of moral value that determines whether cared-fors are justifiably included in care ethics (i.e., ones that have inherent moral value and have the relevant capabilities for participation in the caring relationship). I want to highlight valuing-attitudes in care ethics and somewhat bracket the question of a singular source of inherent moral value for all plant life. Answering that question would be a project in and of itself. There are many different arguments, many mentioned in Chapter 2, about why plants may have moral status (e.g., personhood, subject-of-a-life, autonomy, narrative identity, telos, flourishing, kinship). Care is a highly contextual ethic. At the end of all things, as scientific and moral paradigms shift, the grounding of inherent moral value may shift. There are widely held beliefs about what plant capacities are—and there are so many different types of plants with varied abilities. Depending on one's paradigm, a carer might think plants are living machines or a part of the divine. So, I am taking a pragmatic approach here in requiring the carer to have attitudes that assume a type of inherent moral value in the cared-for as a way of leaving this way open to verify the correctness of that attitude as paradigms shift and change.

In accepting one of the positions in the previous chapter on plant moral value as the ground floor, there is space for carers to have more robust sources of moral value for their plant cared-fors that other moral agents may not accept. For example, in the case of a bonsai practitioner who considers her bonsai trees to be persons (see Chapter 5), her love for the trees suggests that the cared-for have inherent moral value. But her particular belief set supports those trees being moral persons. While a skeptic might not be willing to agree that the trees have such a robust set of morally relevant capacities, she may be willing to accept biocentric arguments; thereby finding the bonsai practitioner's attitude of love coherent rather than absurd or mistaken. The requirement of valuing-attitudes rather than a singular acceptable account of the source of the cared-for's value makes this normative structure possible while also honoring a wide spectrum of beliefs about plants amongst plant practitioners and carers.

Caring Affect

Affect is central to human experience. Feelings can motivate, inform, and direct attention (Little 1995). One of care ethics' major contributions to contemporary metaethics and normative ethics is highlighting morally useful feelings (Palmer 2010, 51–4). Rather than denying that affect has any place in ethical decision making, care ethicists maintain that certain types of affect can actually be beneficial to understanding morally-relevant cared-fors and their needs. When one thinks of the moral relationship between mother and child, feeling is in the background feeding the relationship, motivating the moral agent (mother) through trying times, and creating a bond between the caregiver and cared-for.

As Margaret Little argues, some care as affect might be useful in determining what is morally relevant or valuable in a given situation (Little 1995, 123). The moral feelings that arise in humans for plant cared-fors can be such valuable tools for navigating the world (Noddings 2013, 131). Caring affect both directs my attention to the needs of the plants in my life and motivates me to respond to their needs accordingly. This affect can be interrogated to determine whether it is indeed picking out a morally salient feature of the relationship with the plant or simply mistaken in attributing moral value to a plant cared-for.

For care ethics, the affect should be of a morally relevant quality. For Noddings, joy is an example of affect that accompanies care (Noddings 2013, 132–47). Joy is a feeling that arises from a reflection on the relatedness of humans with other humans and life (Noddings 2013, 132). Noddings even includes other living things in the relatedness that produces the affect of joy. So, this affect can be present regardless of the status of personhood. Perhaps some cases (my great grandmother's jade, the family's olive grove, etc.) can be tied to human family legacies, or be understood as an extension of human identity or ownership. The affection can reflect the appreciation that arises out of a long-term relationship with another being. Sometimes this affect emerges from caring relationships that stretch over human and plant generations (e.g., individual and groups of olive trees in the grove) or emerges from the experience of kinship. The feeling is a response to something of value.

Care can motivate the caregiver. It can add context to the relationship, draw the caregiver's attention to relevant features of the plant and its state that inform the caregiver's actions. So, affect can and does play a role in human-plant relationships. Within the case studies in Chapters 5, 6, and 7, I will provide concrete examples of what this looks like and how it functions in ethical care.

Dependency

A major contribution that care ethics has made in the larger ethical literature is the centrality and moral dimensions of dependency. Humans are fundamentally dependent organisms. We need support in a multitude of ways from the moment we emerge from our biological (m)others to our final moments. The human condition is one of dependency on our environment and our communities. Care ethics incorporate this phenomenon of dependency into the core of ethical theory. But in a care ethic that includes plant cared-fors, how can dependency be included? The dependency in care ethics addresses both physically dependent relationships (e.g., parent and child, patient and physician, elder and caretaker) and socially dependent relationships (e.g., friends and lovers).

Physical Dependence Relationships

Humans are physically dependent on plants. Not every human is dependent on every plant, but for living and flourishing, human life requires ingesting plant matter for sustenance and medicine. Plants are used for shelter. The generic historical dependency of human life on plant life is plain. But particular relationships of dependency between individual humans and individual plants are deeply context dependent and vary widely.

Many plants are thought to be generically independent of human life from a strictly scientific position. Humans could come or go, exist or cease, and it would not make much difference to plants unless either human actions intervened with plant life or plant life was developed by and depended on human intervention (e.g., genetically modified crops, non-native non-invasive species, etc.) Human "footprints" have become so far reaching, however, that in some ways plant life is dependent on humans *to do nothing* (e.g., to cease expanding into and disrupting ecosystems, to stop using fossil fuels and emitting GHGs contributing to climate change, to stop infringing on plant lands and changing their habitats too rapidly), or, in other cases, dependent on humans to restore or repair damage done. Still, even this assertion of independence can be controversial depending on one's worldview. Humans are members of the biotic community (Leopold 1949). As members of that community, our actions ripple into the lives of other beings, even those we may not consider ourselves to be in strict causal relationships with. As will be discussed more in Chapter 7, even "wild" plants have evolved in contact and interaction with human beings, in many cases primarily through fire (Hay 2025). Parsing apart these long ranging historical, developmental, causal, and cultural interactions and inter-dependencies can lead to a difficulty in putting forth blanket statements about independence or dependence between humans and plants. This is where critical plant studies and the plant humanities are so crucial for analyzing these histories, narratives, and practices around human and plant relationships.

One example emerging from indigenous studies is the moral obligation of humans to interact with the natural world. Within some native, indigenous, and First Nations communities, song is both a gift and moral responsibility that humans are required to provide for the natural world (Molina and Shorter 2025). This signifies a largely human responsibility that the plants and other kin rely upon for maintained balance (Anderson 2005). Without these acts towards the natural world, these traditions hold that things will fall out of balance and many will suffer, humans and plants included.

Ultimately, care depends on context. Thinking about myself as a moral agent and my plant-community-context, the solitary purple and green plant sitting on my desk, gifted to me by my partner, requires different action on my part than the tree outside my window or even the sage in the window box. Care is contextual and relies on an understanding of the relationship that the moral agent has with the plant in order to know what one should do. And my very survival depends on the plants in my life: the wheat in my bagel, the tomato and greens in my salad, the soy in my vegan patties. Even the consumption of salmon steaks, beef burgers, or pork bacon would be dependent on plant life supporting and nurturing those animal bodies to the point where they are slaughtered for human sustenance. This dependence might be considered a case for the instrumental rather than inherent moral value of the plants in our lives, since they are required for the sustenance of our human lives. But if the thing we are dependent on is morally valuable in and of itself, then this strict instrumentalization should be resisted. Though a human baby may rely on the services provided by human caretakers, the caretakers are not merely services—they are beings with inherent moral value, who resist pure instrumentalization. Perhaps the same is true of the myriad plants humans rely on for our survival.

Social Dependence Relationships

In the human context of care ethics, physical care (e.g., providing milk for an infant, or warmth for a child, or, more indirectly in our capitalistically structured society, working forty hours a week to have the purchasing power to obtain food, housing, and other material needs for one's human family) is a fundamental and necessary part of care. But there are other senses of dependence in care ethics. Some caring relationships are more socially dependent, like friendships, romantic or loving partnerships, and parent-child relationships after a child is no longer dependent on the parent for her physical needs.

The cultural phenomenon of plant parenthood is a good example of this social dependence in the context of relationships between humans and plants. Humans who consider themselves to be plant parents are dependent on having plant cared-fors for their identity. The caring relationship contributes to their relational selves, comprising one of the primary relationship types that make the carer who they are. Humans who work with plants in caring relationships define not only their work but sometimes their very identity through these relationships.

As Noddings notes in her discussion of reciprocal dependence between the carer and cared-for.

> Clearly, the cared-for depends upon the one-caring. But the one-caring is also oddly dependent upon the cared-for. If the demands of the cared-for become too great or if they are delivered ungraciously, the one-caring may become resentful and, pushed hard enough, may withdraw her caring. Each of us is dependent upon the other in caring and moral relationships.
>
> (Noddings 2013, 48)

For care ethics to function, there is a requirement from both carers and cared-fors to contribute to the relationship. Kinship claims can respond to the demands of social dependence well, I think. If entities are related, there is some connectivity between them that is social and identity based at its core. Who I am is defined in large part by who I am related to. If plants are my relations, then they are part of what comprise the story of who I am. So, there is a social dependency for the carer on the plant cared-for and their contribution to the caring relationship through responsive interactivity and success conditions which will be discussed below.

Responsive Interactivity

Noddings and Held term this component of care "reciprocity." Kittay describes it as "responsiveness." As you'll see in what follows, I prefer "responsive interactivity" in the context of caring relationships with the more-than-human community. I think this is an issue of meanings associated with and affirmed by the terms reciprocity and responsiveness. Reciprocity in Noddings' account is incredibly robust, and includes rich exchange requiring extensive cognitive and emotional capabilities. As what now sounds like an all too familiar refrain, the capabilities individuals are willing to grant to plants will designate whether plants are truly able to be in reciprocal relationships with humans. An example of this within this project is the choice I've made not to consider plant carers. Were plants able to be carers (a position I hope to explore more seriously in future work), reciprocity might be a less controversial claim. However, if plants are incapable of one day being carers, then reciprocity begins to look less possible. Reciprocity might infer a type of give-and-take that is more clearly apparent in relationships between beings that can both be carers and cared-fors. With other humans, we assume intentionality, choice, affect, and concern on the part of a carer, such that reciprocity might work more seamlessly within a human-only group.

Lori Gruen offers an excellent example of reciprocity with other animals in an experience with a chimp recounted in *Entangled Empathy*. Upon entering an enclosure with a new-to-her chimpanzee, Gruen was unaware that her trepidation over meeting a new chimp had raised her heart rate. Encountering Gruen, the chimp pressed her chest to Gruen's, an attempt at slowing Gruen's elevated heart rate, putting it in synch with the chimp's (Gruen 2015). This type of activity demonstrates a type of reciprocal care. The chimpanzee recognizes something off about Gruen's state and attempts to remedy it. I contend that my dog, Sprite, does this frequently as well. One reason I am unable to play video games in his presence is that he senses my heightened agitation with the game (I am too intense) and attempts to get my attention and distract me from the stress inducing stimulation. In relationships like these with other animals, it is easier to identify a type of reciprocity than with plants, because other animals can be quite similar to humans in the way we respond to one another in our animal ways with our animal being.

Some work in critical plant studies more forwardly asserts reciprocity in the relationships between humans and plants. For example, Robin Wall Kimmerer (2013) understands right relationships to be ones of gifts—when we acknowledge the gifts of nature (and plants) and reciprocate by giving gifts to nature. Reciprocity is challenging in this context, because the meaning of reciprocity may depend on what exchange or services one is giving the other. When I think of my relationship with my mother as reciprocal, there is much emotional labor that we provide one another when we are in need. I've also been able to be there for her when she is going through medical issues and she is there for me navigating illnesses. When I think about my relationship with plants, some can offer support during times of ailments (e.g., consuming sage when I have a sore throat) and I can offer aid when the plant is under attack from insects. But some might observe a lack of emotional interaction that is easier for some to experience (or project) in relationships between animals (humans included). So, some claim reciprocity between humans and plants straightforwardly while others consider reciprocity to be something only possible between humans.

However, because in this project I am especially sensitive to more conservative accounts of plant capabilities, I want to be more cautious than some of my colleagues in attributing reciprocity to human-plant relationships. The ability to communicate, for example, might seem critical to reciprocity and a capability that some will not ascribe to plant life. This can be especially true for humans who are reticent to equate chemical signaling with communication. When a plant releases a chemical indicating it is under attack by insects or herbivores, there is a wide range of interpretation about what is actually happening. Is the plant providing information for other plants in the area to fortify their defenses that constitutes calling or crying out? This connects with discussions in the previous chapter on anthropomorphism and meaning in descriptions of plant activity. For some reciprocity is a non-starter for relationships with other animals, plants, and even in some cases, with some other humans.

As one can see from Kittay's preference for "responsiveness," not all humans within that moral group are capable, or will ever be capable, of a robust, cognitively abstract notion of reciprocity. In responsiveness, the cared-for can respond to actions from the care-er, but I prefer using the term interactivity because it emphasizes the relational connection between carer and cared-for. Also, skeptics who consider plants to be non-sentient may be more accepting of interactivity. Regardless of terminology, the caring relationship requires both carer and cared-for to be responsive, both parties to shift and change, both parties are oriented in a certain way towards the relationship.

Mark Coeckelbergh's relational plant ethics captures a facet of this requirement. He describes human and plant interactions as a type of co-created experience (Coeckelbergh 2019). Humans contribute to the experience, but so do plants, through their responsiveness to human activities of watering, planting, pruning, and even destroying. Without the existence and active responses on the part of the plant, the experience of the human-plant interaction would not exist, a relationship compatible with care ethics would not be present between the human and the plant. What makes a plant different than other things, for Coeckelbergh, is not the fact that it is living (as he argues that his analysis also applies in the case of robots and other mechanical human creations), but rather the activity of the plant (Coeckelbergh 2019, 99–100). The plant is able to move, to shift, to emit chemicals, to grow, to age: and in all these ways, it responds to the relationship it has with humans (and even, in some cases, other animals and plant life) in its communities. The plant participates in its community, and in its relationship with humans.

In many human cases, even if two humans are engaged in a caring relationship, only one of them might be able to provide care in a given context. In a state of morphine-induced dementia, my grandmother cannot provide care for herself, let alone my grandfather, my father, me, or her nursing aide. If I provide care for her in that state, knowing that she cannot reciprocate in the moment, and will never be able to again, I do not disqualify our relationship from being a caring one. If she becomes comatose and is unable to respond to my actions of care verbally or behaviorally, that does not disqualify me from giving her care. As she is still interactive with its environment, medicine, and my actions, even in her dying.

A key component of an ethic of care inclusive of plants and in contexts where the capabilities of cared-for are contentious is to unseat the reciprocity requirement in favor of an interactive requirement. A plant can still communicate success or failure of a care action by coupling responsive interactivity and the final component: success conditions.

Success Conditions

How caring actions are received by the cared-for is essential to care ethics. Noddings discusses the completion of care that requires the acceptance of care given by the carer to the cared-for by the cared-for (Noddings 2013, 37). Kittay uses the example of a plant when explaining the necessity for success conditions for care,

> Let us begin by considering a very simple example: caring for a plant by watering it. Now suppose that unbeknownst to me, I pick up a glass of *vinegar* instead of a glass of water to pour into a dry wilting plant. Predictably, the plant begins to wither instead of perk up. Have I cared for the plant? Most of us would answer, "No."
>
> (Kittay 2019, 187)

The plant's response in this case (withering when the caregiver mistakenly gives it vinegar instead of water) indicates a failure of success in the caregiver's attempt to administer care. The plant responds to the caregiver's actions, and that responsiveness informs the caregiver of their success or failure in caring.

The possibility of beneficial outcomes of care indicated by the cared-for not only fulfills a core requirement of the normative framework of care, but also demonstrates that the cared-for must be the type of thing for which there can be beneficial consequences. The way the world goes for the cared-for matters in a way that can benefit its well-being.

Assent to the care from the cared-for is another crucial part of success conditions. Not only does good care contribute positively to the cared-for's wellbeing, but also the cared-for demonstrates the acceptance of the care. The cared-for is responsible in the caring relation to communicate whether they accept the care given by the carer. Maria Schörgenhumer gives an aesthetic account of how plants communicate this to human carers (Schörgenhumer 2019). She contends that the aesthetic beauty that results from plant flourishing (e.g., bright colored leaves, sturdy stems, bold flowers, etc.) is a way that plants communicate their recognition of care. If a plant is not accepting of the care, the plant will have aesthetically disturbing traits like yellowing or splotchy leaves, withering appendages, and misshaped components. While a skeptic might doubt that a plant is actively communicating in a way that engages beauty, surely plants demonstrate the success conditions of care through sensory available-to-humans, or aesthetic, means. Humans observe these changes and make judgments about the plant's health. In this way, like Kittay mentions, the plants have a means of showing their acceptance or rejection of the care given.

The Upshots and Concerns

If one were to accept this account of plant cared-fors and care ethics, there would be a few major outcomes. Beyond acknowledging more morally rich relationships in existent communities, we may be called to establish even more caring relationships with other cared-fors. Admitting plant cared-fors into care ethics also encourages the development of sister virtues. Additionally, a plant-inclusive care ethic could connect local communities with larger global crises, by enriching the moral community at home supporting local changes that can influence the outcomes of global issues. However, these upshots come along with concerns that might arise from the those reticent to include plant cared-fors in care ethics.

Upshot: Cultivating More Caring Relationships

As a normative theory, care ethics requires acknowledgment of existing care relationships and encourages entering into more caring relationships. While Noddings contends that we have a universal obligation to care for other human beings, perhaps we have a general obligation to consider caring for all potential cared-fors. For example, when I consider only relationships with person cared-fors to be morally considerable, my immediate circle of cared-fors includes my human family members, my spouse, my cat, and my dog. Of course, caring relations also extend to the other persons in my life like my students, employers, supervisors, colleagues, friends, the pets of my friends. When I open up potential cared-fors to include plants, I instantly more than double the number of cared-fors in my immediate surroundings because of my houseplants. They outnumber the animals in our household. If you extend this to cared-fors on our (rental) "property" a catalpa tree and numerous perennials are added to that group.

I see this as a beneficial outcome to this ethic because it requires us to take seriously the relationships we have with cared-fors in our immediate environment. It grounds us in these existing relationships and illuminates the investment that we must make where we must provide successful care to those in our communities as well as enhances our awareness about the benefits we receive in caring relationships with them. As we realize the greater number of morally rich relationships we have in our immediate communities, our interest in developing more caring relationships with other cared-fors will likely grow.

We can start to realize the relationships we have to the foods we eat and the general consumption patterns we form. Acknowledging the interdependency, we have as carers with the growing number of cared-fors will require us to move more carefully through the world.

Moral Consideration: Obligations to Care for Too Many Things

One major concern that arises from admitting a larger number of cared-fors into care ethics is the limit of what is possible. If I simply name the possible cared-fors in my environment as I write this, there are at least 16 house plants of various species, my dog, my cat, my partner, the asters outside, the catalpa tree, English ivy, myriad bushes and unknown-to-me outdoor plants, squirrels, nuthatches, and woodpeckers. How am I supposed to care for all of these creatures? I would probably never get sleep if I were attending to their every preference the way that my partner and I attend to our cat and dog.

This comes from a mis-construal of care ethics, I think. There are two points that are important here: first, that obligations may differ towards different cared-fors; and second, being a potential cared-for does not mean that care must be provided at all times, in all places. To the former, I will begin with the wide variety of relationships I have within my human community. The care obligations that I have for my parents are different from the care obligations I have towards my students. My spouse requires a different type of care than nephews and nieces. My care obligations to my immediate family (spouse and pets) are currently more frequent and time consuming than my obligations to my parents and brother, though when I was younger, and I lived with my parents and brother (and didn't have a spouse), the pendulum swung towards family. The care I might be required to provide for humans in my local community also varies greatly. If I were to happen upon someone collapsed on the sidewalk, I may have previously not give them any care, but given their current condition and the fact that I am aware of them and in a position to provide care, my obligation to them might shift depending on our context and relation.

Next, just because something is a potential cared-for does not require an obligation from every carer to give it care. The context of the caring relationship and the identity of the carer and cared-for are important. For example, my dog Sprite is (I think uncontroversially) a cared-for. He has inherent moral value and capabilities that make it possible for him to be a cared-for. Does that mean that he is a cared-for in an ethically caring relationship with Alba, a human moral agent in Bismarck, North Dakota, who has never met and will never meet Sprite? Of course not. Ethical care is found in a relationship between a carer and cared for. The same could be said of Alba's proposed obligation to the angel wing begonia on my front porch. However, there are plants that Alba is a position to care for that she would be obligated to care for, even if she has no house plants or owns no land on which plants grow.

Even when looking at care ethics within human communities, I do not provide care in the same way to every person I come into contact with. In fact, I may not be required to take a caring role for every person on earth. Rather, care ethics calls us to attend to our caring relationships, as they form from the mass of relationships with have with others. Care might allow for non-intervention with others at times, or more invasive support of cared-fors at other times. Not only does the context have importance in determining the contours of care given, but it also can pick out the relationships that require high intervention care at a given time.

The caring relationship is context rich. For a carer to successfully provide care to a cared-for, they must be attentive to the identity and needs of the cared-for. The way I would provide care—or even be able to provide care, given that we are renters—to the catalpa tree outside my window is very different than the angel wing begonia in my living room or the jade plant beside me or our cat Aria sitting in the chair next to me. The context of the caring relationship matters because it is about the carer and cared-for. Both as finite biological organisms, there are inherent limits of services provided and received. It is up to the carer and the cared-for to navigate these finite realities.

Sandler addresses a similar concern that critics might have in accepting that plants have inherent moral value. He states:

> There are many plants. However, there are also many people. A multitude of a kind of thing is not itself sufficient for justifying the claim that it is not possible to take the interests of individuals of that kind into account. (Sandler 2019, 48)

In supporting this response, he also notes that the challenges to consider all of the interests of a great number of people for a given normative ethical theory is not unlike the challenge of considering an even larger number of beings with inherent moral value.

With this in mind, the concern might be less about the inclusion of more cared-fors and more about the more general structure of care ethics and how care ethics as a normative theory can account for these numbers compared to other normative ethical theories. For the time being, it may suffice to say that this particular concern would be outside the purview of this book, which is aimed at making sense of how plants can be included in care ethics rather than evaluating the way that care ethics can stand up against other normative ethical theories when it comes to large numbers of entities with moral considerability.

Moral Value Hierarchy: Discriminating Amongst Cared-fors

Another concern that might come from allowing plant cared-fors into care ethics might be the theoretical and practical concern of all cared-fors being of equal moral value. This is a concern that Noddings and Kittay raise when considering whether other animals could be cared-fors in care ethics. They found the idea of a human infant having the same moral value as a dog or a rat wrong. When all your cared-fors are human persons, it may be easier to avoid investigating whether all cared-fors have equal moral value.

So, if we admit that not only dogs and rats can be cared-fors, but also begonias and catalpa trees, an interlocutor might find this not only absurd but offensive. Surely my human nephew is worth more than my begonia. So, if plant cared-fors have the same type of moral value (inherent) and moral considerability (cared-for) as human persons, then it should be rejected outright.

But what we might be getting wrong here is the assumption that having inherent moral value and the moral consideration of a cared-for equate to the same interests and treatment (Sandler 2019, 45). As Sandler notes, plants have different interests in light of being different types of organisms and would warrant different treatment in light of what matters to them. The needs of my dog are different than the needs of my begonia relevant to their very different interests, the former's psychological and emotional social needs and the latter's need for sunlight. My dog is a different type of creature, and so the attentive care that I give him is responsive to his needs and desires for flourishing. The same goes for my begonia and its unique needs.

There is also space beyond the threshold of inherent moral value and distinction of cared-for to have different hierarchies of care (Kagan 2019). Even considering other human cared-fors, I may have different moral obligations of care to my (hypothetical) own human child (regardless of whether she shares my DNA) than to one of her classmates or her cousin who lives several states away. When we shift to consider more-than-human cared-fors, similar relational contours pick out our particular obligations of care. My houseplants are dependent on me for care in a different way than the wild plants on our rental property or at the local nature preserve or even the local farms where we source our vegetables. The relationships between us and the character of the cared-fors determine what is required by the carer. And so, the nuances of caring affect, dependency and responsive interactivity demonstrate how the caring relationship can vary from one cared-for to another.

Action-Guiding: Dilemmas Between Different Cared-fors

With all of these various cared-fors (human, animal and plant), there are going to be dilemmas between providing good care for all of the cared-fors. Especially when other animals and plants are included as cared-fors, the issue of consumption is ever present. How do we decide if we can kill one cared-for for the well-being of another? And how universalizable is this care ethic?

One response to this important concern follows from the above: having inherent moral value may not mean that it is a Reganian sense of inherent moral value. That is, for example, that the being with inherent moral value may never be killed. Even in Regan there are some spare examples of when the being with inherent moral value's life may be overridden. For example, Regan offers such instances in his mini-rider principles that explain why in a lifeboat situation a dog rather than a human may be killed (Regan 1983, 285–6). But when plants have inherent moral value and are cared-fors, there will certainly be more dilemmas than we face even when looking at human and other animal cared-fors. For example, it is possible for many humans to physically survive without consuming animal products. But this is not the case for plant products. What makes sacrificing a plant cared-for okay and sacrificing a human cared-for not okay?

As mentioned above, there can be differentiations between entities with inherent moral value. This can be a first step to figuring out how to interpret the contexts of the caring relationships to determine what is best in a given situation. But one might worry that the plant cared-for might get overridden so many times in favor of the human and animal cared-for's wellbeing that it seems as though having inherent moral value does not practically matter.

This is where care ethics is particularly distinctive as a normative theory, I think. Lori Gruen discusses this distinction as care ethics' move away from other ethical theories that concentrate on abstraction and universalization to a particularist ethical project (Gruen 2015, 17–18). The context of the caring relationship and other competing caring relationships between the carer and numerous cared-fors make abstraction to general rules of action-guiding not useful. Care is embedded in real relationships that defy abstract generalizations for rule-guiding obligations. This may be a larger issue with care ethics itself as opposed to other ethical theories. But I do not think the inclusion of plant cared-fors alone adds any particularly menacing complexity to this decision-making process.

Upshot: Proliferation of Sister Virtues

Fostering more virtues (i.e., morally good character traits) is one positive outcome from adopting a care ethics framework coherent with plant cared-fors. There are many sister virtues that will likely be cultivated alongside the adoption of this care ethics framework. Other theorists maintain this claim between plant care and virtue development as well. Within the context of the garden, philosopher David Cooper argues that "certain garden-practices necessarily induce virtues" (Cooper 2006, 93). A host of intellectual virtues develop within the caring relationship with plants. A carer must be observant and curious, seek knowledge when she needs direction to bolster the health of the plant. To better understand the plant, the carer may be called to do non-invasive experiments trying to enrich the plant's life.

The carer can also acquire a number of moral virtues. She learns humility when caring actions do not yield the desired outcomes for the plant and cultivates gratitude for the cared-for's presence in her life and in light of any benefits the cared-for provides to her. In caring for plants, the carer increases her awareness as a highly interdependent being. This encourages the development of virtues like wisdom in navigating this existence.

There are also ripples throughout the community in response to the development of virtues in individuals. Beyond the positive mental and physical health benefits for the carer, but it can also provide the community with food and medicine. The cared-for plants give back to the community and encourage the development of gratitude in the carers. These virtues not only benefit the carer by developing better character, but also enhance the community and provide a context for supporting virtue development in others. The virtues practiced by the carer can inspire and have a meaningful impact on other carers.

Action-Guiding: Proliferation of Brother Wrongs

If sister virtues are proliferated in the adoption of plant cared-fors, an interlocutor might argue that the adoption of plant cared-fors also proliferates moral wrongs. Because plants can count as cared-fors, when we kill them to consume them, are we not exponentially expanding the times that we are failing to care? It is useful to think of plants that must be killed in order to be consumed, like beets or onions. How can we consider this care ethic possible if it requires us to regularly act immorally because we must kill some plants to survive, or in less severe cases, we kill them either for taste or important cultural practices. Consumption occurs over a wide range of practices, from treating plants like mere objects whose lives are totally meaningless (e.g., when plants are overharvested, they are used for nothing and result in furthering GHG emissions through waste management operations) to thanking a plant when one takes its life. Given this wide disparity of approaches, clarifying the context of killing is important.

First, we should acknowledge that killing is not always unjustified or may not be in alignment with supporting a cared-for's wellbeing. I know this sounds controversial, so let me explain. There are arguably instances with humans and other animals, for example, where euthanasia is more in keeping with honoring their wellbeing. This is a controversial topic, but with the widespread practice of at least euthanizing other animals for the reason of curtailing needless suffering or a life not-worth-living (again, very controversial), killing an animal with inherent moral value may be in keeping with caring for them.

This context may appear very different from the context in which humans kill plants for food. Often in my contemporary context, food plants are reared in industrial conditions and culled just before or at their height of health as an instrumental purveyor of nutrients to the human consumer. Also, in contemporary agriculture, as we'll see with tomato plants, some plants are reared in contexts that might mean they are annuals rather than perennials. Thus, their lifespans are shorted even if not directly or intentionally killed. Also, many plants are killed for consumption but not consumed or composted, so there is not even any instrumental value for them; they simply become waste. Where is care in this system? It seems clear that it is absent. But that is not an indictment against plant cared-fors, but the systems and institutions that produce and cull them without the presence of care. It is an indication that care is needed to help improve the food situation. There are approaches that favor limiting the deaths of plants based on true needs of humans, in some cases to the point where excess harvesting results in the plant "cursing" the person who overconsumed it (Molina and Shorter 2025).

So, if we imagine an idealized situation involving the beet where it is used for food outside of contemporary practices, wouldn't we still find that this practice requires the undermining of the beet's wellbeing, ensuring that the killing of the beet requires a moral wrong via care ethics? This is a difficult issue since humans and other animals require the killing of plants to survive and flourish. Some in the plant studies claim that plants sacrifice themselves to those who consume them. Were plants able to make these choices, then perhaps they could be considered carers and humans and other animals their cared-fors. However, a skeptic of plant capacities would likely not accept such an account of plant capabilities, death, and consumption.

For the interlocutor, the issue of death would likely be connected to the plant's biological interests and the way that killing a plant undermines them. Given that death is a given for living things, one might be concerned about what is lost when one dies. For the plant, its biological body is changed. Since the skeptic would not grant a psychological life, the biological body is pretty important to the plant. But what does flourishing mean for the plant? It might include reaching a stage of development where it can successfully reproduce, going through the major life stages without being overcome by illness or pests, passing on of genetic material (of etiological significance), and living a long (plant-relative) and healthy life. Some of these things are compatible with killing. A plant might be able to do all these things, live into relative old age, and then be harvested for consumption. It is also possible that these things do not matter to the plant. This is a call for further investigation into what interests the plants have and how to best support their wellbeing and interests alongside the wellbeing and interests of their communities.

Upshot: Local Relationships Foster Global Communities

Caring relationships naturally may start at home, but they can expand to long-distance relationships, as I will address in Chapter 7. If I were to wake up today to all of the ethically laden relationships with plants in my immediate relational and geographical vicinity, it may be surprising how much possibility there is for care to be cultivated. One might think that I would have more than enough on my plate considering all of these close relationships. But attention to and consideration of these caring relationships naturally leads to consideration of other possible caring relationships with other plant cared-fors. By beginning with our local plants, we can orient ourselves to their needs and flourishing in a way that cultivates our interest and commitment to care for more living things in our communities.

But many of the environmental problems that we are confronted with are not restricted to our immediate relational or geographic vicinity. How can a local-attentive care ethic address immediate and impending existential environmental crises? Conventional wisdom tells us to get ourselves into good working order before taking on complex relationships with other humans and to take care of oneself before devoting ourselves to the care and support of others (Savage 2006). We should get our own houses in order. Starting locally does not preclude growing our communities of care beyond the immediate vicinity. Rather, it calls for us to start where we are and build from there. Developing caring capacities at the local level can support further enactment of this care on the global level.

It is also a practical approach. Carers do not need to go out and buy things to cultivate care. They can begin exactly where they are and become better attentive and invested in the plant cared-fors in their own environments. Whether in city, suburban, or rural environments, our daily lives are inundated with plants. On our plates, sidewalks, and fields, plants are ubiquitous. This can produce even greater interest beyond one's own immediate space and expand to understanding and solving larger problems that involve human, animal, and plant cared-fors.

Conclusion: Coherency and Plant Cared-Fors

The discussion above makes the case for plant cared-fors clear. Not only can plants qualify as cared-fors within care ethics, but their inclusion also gets closer to being right about the moral make-up of our multi-species communities. It also makes sense of different cultural practices including the social phenomenon of plant parenthood. While there may be theoretical and practical issues when including plants as cared-fors, I argue that the upshots provided above outweigh those legitimate concerns. When incorporating different beings into our ethics, there will inevitably be points of friction, because the anthropocentric frameworks were not created with more-than-humans in mind.

Accepting plants as cared-fors also maps on to some intuitions about the moral value and capacities of plant life beyond human use of plants as mere means to an end. The tension running throughout this text so far has been one of cultural, philosophical, and scientific difference. Some are willing to grant incredible capabilities to plants, while others are willing to grant relatively few capabilities. I don't want to denigrate the more robust in favor of the more conservative, nor vice versa. I think providing an account that is salient despite this formidable reality of diverse belief sets about plants is important at this time. In the midst of shifting beliefs and worldviews, theory that is flexible without being meaningless or overly reductive is necessary for providing a bridge to future directions and developments.

Now that I have demonstrated that plant cared-fors are coherent within a care ethics framework, it is time to evaluate the theory in real caring relationships. The next three chapters address three distinct caring contexts with plants: bonsai trees, tomato plants, and sequoia trees. By applying care ethics with plant cared-fors to these particular contexts, I can evaluate whether the ethical theory is possible and actual, as well as identify any context-specific danger zones that might undermine good care for plants in the care relationships.

References

Anderson, M. Kat. *Tending the Wild: Native American Knowledge and the Management of California's Natural Resources.* University of California Press, 2005.

Cabrera, Daniela. "How to Become a Plant Parent." *New York Times.* May 14, 2018.

Coeckelbergh, Mark. "What Do We Mean by Relational Ethics? Growing a Relational Approach to the Moral Standing of Plants, Robots and Other Non-Humans." In Angela Kallhoff, Marcello Di Paola and Maria Schörgenhumer (eds), *Plant Ethics: Concepts and Applications* (pp. 98–109). Routledge, 2019.

Cooper, David. *A Philosophy of Gardens.* Oxford University Press, 2006.

Gruen, Lori. *Entangled Empathy: An Alternative Ethic for our Relationships with Animals.* Lantern Books, 2015.

Hay, Elspeth. *Feed Us with Trees: Nuts and the Future of Food.* New Society Publishers, 2025.

Kagan, Shelly. *How to Count Animals, More or Less.* Oxford University Press, 2019.

Kimmerer, Robin Wall. *Braiding Sweetgrass: Indigenous Wisdom, Scientific Knowledge, and the Teachings of Plants.* Milkweed Editions, 2013.

Kittay, Eva Feder. *Learning from My Daughter: Valuing Disabled Minds and Caring That Matters.* Oxford University Press, 2019.

Leopold, Aldo. *A Sand County Almanac.* Oxford University Press, 1949.

Little, Margaret. "Seeing and Caring: The Role of Affect in Feminist Moral Epistemology." *Hypatia*, 10(5) (1995), 117–137.

McShane, Katie. "Why Environmental Ethics Shouldn't Give Up on Intrinsic Value." *Environmental Ethics*, 29(1) (2007), 43–61.

Molina, Felipe S. and David Delgado Shorter. "Episode 47 Felipe S. Molina and David Delgado Shorter." Networking with Plants in the Anthropocene Podcast. November 1, 2025. https://soundcloud.com/networking-with-plants/episode-47-felipe-s-molina-and-david-delgado-shorter

Noddings, Nel. *Caring: A Relational Approach to Ethics & Moral Education*, 2nd edn. University of California Press, 2013.

Palmer, Clare. *Animal Ethics in Context.* Columbia University Press, 2010.

Regan, Tom. *The Case for Animal Rights.* University of California Press, 1983.

Sandler, Ronald. "Is Considering the Interests of Plants Absurd?" In Angela Kallhoff, Marcello Di Paola and Maria Schörgenhumer (eds), *Plant Ethics: Concepts and Applications* (pp. 40–50). Routledge, 2019.

Savage, Dan. "The Savage Lovecast." (2006–present). https://savage.love/

Schörgenhumer, Maria. "Plant Ethics: Concepts and Applications." In Angela Kallhoff, Marcello Di Paola and Maria Schörgenhumer (eds), *Plant Ethics: Concepts and Applications* (pp. 98–109). Routledge, 2019.

4 Cartography

Charting Interdisciplinary Paths in Case Study Research

Introduction

This book is divided into two halves. The first half argues for a care ethic that can include plant cared-fors, while the second half engages in normative case studies (Taylor 2024). These case studies navigate the rugged contours of caring contexts in actual relationships between humans and plants. In keeping with the spirit of environmental ethics, widely considered within the realm of applied ethics, I draw from theory and practice (Hourdequin 2015). Given this shift between theory and practice, it is worth dedicating space to this transition and making clear what the case studies are and what they are not. Improperly understood and contextualized, this section of the project could appear to be an ill-executed inadequate attempt at social science research. This is particularly true if one is expecting three chapters of sociological or anthropological qualitative case studies. That would be a mistake, though a natural one to make. I am working in a slightly different area than qualitative case studies, that of normative case studies. Rebecca Taylor offers a solid account of normative case studies "as a form of empirically engaged philosophy, in which case studies represent empirical phenomena in ways that highlight normative questions that emerge in those real-world contexts" (Taylor 2024, 303). Not only does care ethics emerge from reflection on actual human lives and relationships, the application of the ethical theory to real life practices can reveal important spaces of further ethical questions and reflection.

Contested Territory: Discussion of Experimental Philosophy, Qualitative Case Studies, Ethnography, and Normative Case Studies

The spaces between the disciplines of philosophy and other disciplines are constantly changing, re-negotiating, and evolving. To better understand where and how this project is situated within philosophy and how it is different from qualitative social science, the distinctions between Experimental Philosophy (XPhi), qualitative case studies (QCS), ethnography, and normative case studies (NCS) are useful. XPhi and NCS are considered empirically engaged

DOI: 10.4324/9781003603665-5

philosophy, while QCS and ethnography are considered social science. By understanding some of the boundaries, hopefully the goals and methods of the case studies in this book can be clarified.

Experimental philosophy (XPhi) is an ever expanding (and sometimes controversial) sub-field of philosophy that incorporates methodologies from the sciences to better understand the relevance of empirical research for interrogating philosophical questions. Archetypical XPhi particularly utilizes cognitive science and psychology. However, XPhi has a range of research topics and methodologies. One thing that distinguishes XPhi is the necessity of philosophers engaged with XPhi having a high level of literacy and competency in psychological and cognitive scientific research methods, which they often employ to draw conclusions about things like moral intuitions.

And so, what unites XPhi as a sub-field and is latter half of the monograph a work of XPhi? As Joshua Knobe and Shaun Nicols explain: "research in experimental philosophy brings together two key elements: (a) the kinds of questions and theoretical frameworks traditionally associated with philosophy; (b) the kinds of experimental methods traditionally associated with psychology and cognitive science" (Knobe and Nicols 2017). While this project satisfies (a) of Knobe and Nicols' XPhi definition, it does not meet (b). An XPhi project on the attitude of plant caretakers might focus on the effects of gender on the care takers' value judgments about plants. For example, during my interviews with bonsai practitioners, I began to wonder if and how gender might play a role in the practitioner's comfort with more invasive limb removal and management practices. If I were to conduct a psychological study of bonsai plant caretakers to determine the effect of practitioner gender on their moral intuitions about human intervention with plant life, that might be a better fit for XPhil. I would be engaging a question embedded in moral philosophy and feminist philosophy and conducting an experiment using psychological research methods to determine whether empirically there is a difference in attitudes related to gender. As you'll find in what follows, that was not what I pursued in this project. The research methods that I utilized were used in service to understand a given context and case study, rather than understand the source of moral intuitions themselves.

So, if this project is not XPhi, perhaps it is based instead in the social sciences more broadly. Given the centrality of case studies in the chapters that follow, one might think that they are practice of qualitative research prevalent in the social sciences. For example, I did conduct interviews with practitioners to better understand how they understood their relationships with the plants they care for. Given these research practices, it might appear that the last half of this project is comprised of qualitative case studies (QCS). However, a quick overview of the differences between QCS and normative case studies (NCS) will show that the focus of these chapters is NCS, not QCS. I think the most attractive candidate for QCS in this project is ethnography, so let's begin there.

Ethnography is qualitative research with a history that includes wide variation (Davis and Craven 2016, 9). Typically used in anthropology, "ethnography typically involves long-term interactions through participation-observation" (10). Specific research methods vary, but may include interviews. In some ways, the case studies provided in this book may appear to be an attempt at ethnography, providing an objective account of current practices of plant care in a particular place and time. However, ethnography usually takes place over a longer period of time to ascertain the truth of the observed individuals' culture and activities, while my interviews focused on beliefs and practices that relate to ethical issues and usually occurred within one session (10). Additionally, I was not observing a group of practitioners in their community and practice with the goal of ascertaining a purely descriptive account of their activities. Some empirical contours of their practices and beliefs are important for developing the real-life portion of the case studies. But the overriding utilization of the interviews in my methodology are in service to making sense of an ethical theory when it is applied to the real world. Especially given this context, the attraction of calling these case studies a type of ethnography is understandable, given that "ethnography is as much about bridging theory and practice—praxis—grounded in an effort to explore situations from the perspective of those living the experiences researched" (10).

The interest in meeting theory with practice is at the foundation of this project. But the goals of applied ethics are different from those of ethnography and QCS.

The aims of NCS are centered on normative ends (e.g., what someone or some group or something should or should not do), and subsequently are evaluated through ethical reflection and argumentation. According to Taylor, this contrasts with QCS, which aim "to answer a broader array of research questions which tend to align with those also used in other qualitative research methods—e.g., exploratory, descriptive, or explanatory questions" (Taylor 2024, 307). While I am interested in what plant care takers on the ground actually belief, my inquiry does not stop there. I want to understand the moral and normative themes that emerge and examine how an ethicist could make sense of these tensions beyond simply providing an empirical description of care practices in a given place at a given time.

Another way NCS can be differentiated from QCS and philosophical thought experiments by addressing their claims on two vectors: complexity and fictionalization (Taylor 2024, 304–5). According to Taylor, both normative case studies and qualitative case studies appear in the same quadrant when mapped: they both have a high degree of complexity and aim towards a high degree of "pure" reality. But the reality of the QCS is descriptive in aim: what are the given practices, beliefs, and characteristics of care in a given place? The reality that NCS aim towards is moral, and concerned with what can be done when facing the descriptive reality of a case study and managing the moral complexity and problems that emerge given that empirical state of affairs. So, NCS appears on Taylor's map a bit more fictional that QCS, because the goal is not merely to provide the facts on the ground about a given practice or situation. But NCS are not so hypothetical that they create the highly fictionalized examples present in philosophical thought experiments. NCS aim to provide a true context that examines not just empirical facts, but the moral issues that emerge from those facts in a given realistic context.

As a philosopher, I aim to approach the following chapters fundamentally as philosophical normative case studies, not social science ones. So, while I have conducted interviews and detail the methodology below in terms common to the social sciences and QCS (e.g., snowball sampling), the aim of the research is aligned with the normative character of NCS. I am exploring a normative puzzle: can humans care for plants, in a moral sense? And in each case, different contours challenge the theory of care developed in the first half of this monograph. The use of case studies here enables me, as an applied ethicist, to explore both anticipated and unexpected ethical issues that emerge from a close study of carers' attitudes and practices in their relationships with plants. As will be illuminated, each case study reveals unique tensions and potential pitfalls for an ethic of care that includes plant cared-fors. In focusing on contemporary practices of bonsai, for example, special problems of hyper-dependency and the possibility of the plant cared-for flourishing arise along with particularly strong claims about plant care takers considering their bonsai plants as kin-like. Given the existence of these challenges, what can care ethics look like on the ground (or in the garden)?

The Recent Practice and Promise of NCS in Applied Care Ethics

In a way, care ethics is especially well suited to take on this case study and methodology puzzle. Historically and theoretically, care ethics has developed engaged with empirical research and takes seriously the context of care. As mentioned earlier in this book, the work of Carol Gilligan, that many identify as the foundation of contemporary care ethics literature, began not in philosophy, but psychology (Gilligan 1982). In providing a place for empirical research within the theory, case studies of care in practice are a natural extension of the care ethics literature. Care ethics also has a rich and growing literature involving case studies and application in interdisciplinary contexts. The case studies highlighted here provide a range of approaches to care ethics NCS. While case studies occur widely in applied ethics, the focus of this survey in care NCS methodologies is in environmental ethics (i.e., ethics focusing on animals, plants, and ecosystems).

Beginning with the other animal literature,[1] the following three case studies offer a range of methodological approaches. Raymond Anthony (2012) uses a theory-heavy approach, while Merisa Thompson (2022) provides a NCS that also qualifies as a QCS. Anthony (2012) merges environmental virtue ethics and care ethics to interpret and evaluate current agricultural practices with farmed animals. His methodology is very theory centered, one might question whether "Building a Sustainable Future for Animal Agriculture: An Environmental Virtue Ethic of Care Approach within the Philosophy of Technology" even qualifies as an NCS. However, I include this article because I understand the foundation of NCS is the use of ethical theories to interpret a particular real-life practice (or set of practices) supported by empirical research. So, this might be very loosely considered an NCS, as the methodology is not based on interviews or surveys, but rather on applying a theory to a particular case.

In stark contrast with Anthony is Merisa Thompson's (2022) application of care ethics à la Tronto and Puig de la Bellacasa to the mother-calf housing conditions of dairy cattle. In this case study, Thompson explicitly describes the case study as ethnography and follows a methodology that could qualify the project as a QCS. The reason I believe it also counts as a NCS is due to its normative nature: "an ethnographic and qualitative approach was deployed that sought to examine 'ethics' as situated and contextual, rather than universal" (Thompson 2022, 697). The normative content and aim of the inquiry could distinguish NCS from QCS, or at least qualify what would typically be considered QCS as also a candidate for NCS. Thompson interviewed a variety of individuals working at five dairy farms that are working to improve animal welfare conditions and existing on the fringes of commercial dairies.

Alternatively, Lissy Goralnik, Matt Ferkany, Laurie Thorp, and Kyle Powys Whyte (2014) provide a NCS on the spectrum between Anthony and Thompson. Goralnik et al. (2014) use an interdisciplinary team to apply Nel Nodding's care ethic in education to the Student Organic Farm at Michigan State University. Working with existing institutions, they situate care ethics both as a central value to the Student Organic Farm, but also in moral pedagogy at the center of their project with the farm. Working dialogically with different parts of the Michigan State institution alongside students and outside institutions, the interdisciplinary group aimed to give students options of raising pigs in different contexts that were in better alignment with care ethics. This type of normative case study is explicitly interdisciplinary, utilizing faculty and research methods in the social sciences to inform their analysis of care on the ground and to determine whether the husbandry and animal care methods aligned with the values and aims of the care ethical framework. Overall, an emphasis was placed on using care ethics to enrich the relationships between students, animals, and others in the ecological and institutional community to develop virtues and create more successful outcomes for the animals involved.

In other cases, ecosystems are the focus. Esteve Giraud (2021) references a long list of care theories, from Gilligan and Noddings, to Tronto, Cuomo, and Puig de la Bellacasa in his analysis of urban permaculture movements. Margaux Alarcon and Pascal Marty (2023) use the theoretical lenses of Tronto and other French theorists to apply care to farms in Wellonia, Belgium. Finally, Seymour and Connelly (2023) apply Tronto, Krzywoszynska, and Whyte and Cuomo to regenerative agriculture in Aotearoa, New Zealand.

Still other case studies examine particular plants. Anna Krzywoszynska (2016) details her caring relationship with vines in Colli Verdi, Italy, using autoethnography and the theories of Tronto and Mol. She analyzes the testimony of farmers and her own experience to make sense of the expertise and attention required in caring for grapevines through pruning. Another theorist using the care ethics of Tronto along with Puig de la Bellacasa, Jose Cañada (2024) analyses Posidonia in Balearic Islands. He argues that Posidonia becomes embedded in culture through narrative engagement, even going so far as to describe the plant as an agent of care. Both of these case studies provide examples of how care ethics can engage closely with and alongside the empirical in applied care ethics and environmental ethics. While Krzywoszynska engages in a more embedded approach (autoethnographic) than I do here, her theory-rich examination provides good cause for considering her work to be an NCS. Cañada's approach is more akin to parts of the NCS that I embark on, though his analysis is more removed from the particular carers in relation to Posidonia. In what follows, I provide my methodology and explain how it qualifies the case studies in this work as NCS.

Methodology

A central concern of any case study, normative or qualitative, is methodology. As Taylor argues,

> NCS should (1) reflect realistic phenomena that raise complex normative philosophical questions, and thus, (2) provide a foundation for empirically grounded normative analysis and theorizing. These basic elements point to some minimal criteria for constructing NCS, while living ample room for diversity in case study forms and in approaches to developing cast studies.
>
> (Taylor 2024, 303)

The application of theory to case studies requires some sense of how things are on the ground. To better understand the state of things, I began with one of the philosopher's most treasured resources: books. I searched plant care manuals for language that might clearly express the author's views on plants. When I say "the author's view on plants," I was looking for their valuation of plants. Many manuals read as basic how-tos, not clearly expressing the authors' more privately held views on plants. These private views were essential to understanding the instantiation of care. Do the authors value the plants in and of themselves, or merely as a means to some end? Have these attitudes of valuation changed over time? Are these valuations of plants universal across plant life, or do some plants matter to the authors in different ways than others? This is the crux of the information I sought from interviews on the ground. It is well and good to have an instruction manual for rearing plants, but to know whether my theory could adhere in actual caring contexts (or contextually possible caring relationships), I needed more refined information that was unlikely to come from a plant care manual.

So, my search led me further into historical research, primary resources like John Muir's reflections on the plants he encountered in the western United

States, and secondary and tertiary resources that documented the history of human interactions with different plants. Some of these texts provided valuable information about shifting attitudes towards some plants throughout historical movements. However, it was still difficult to get beyond this general understanding of a specific set of plants to the attitudes and values of particular carers. In some cases, especially in indigenous histories, plants may have a specified value or role within stories. However, in many non-indigenous contexts, plants were backgrounded or commodified. There is a well-documented phenomenon of plant awareness disparity, formerly known as plant blindness (Parsley 2020). I cannot help but wonder if this awareness disparity is part of why plants are encountered as mere resources or manipulatable green machines for human use in so many cases. So much so that it isn't even worth explicitly addressing this as their value.

Due to a growing contemporary interdisciplinary area known as critical plant studies or the plant humanities,[2] plant keeper memoirs and plant narratives are also available now (e.g., Wohlleben 2015, Gagliano 2018, Gibson 2024). These works more explicitly identify the author's moral and spiritual views of plants, along with delineating capabilities that they ascribe to plants, like autonomy, agency, and sentience. Many of these documents, however, do not address directly the plants and practices that I have chosen for this project. So, while they are incorporated, I still needed to talk with actual plant caretakers about their valuing attitudes towards their plants for developing my normative case studies.

Interviews with Plant Care Takers

I began informal interviews with plant care takers in the summer of 2021. These interviews extended through the spring of 2023. There were distinct lists of questions I provided each interviewee, depending on the case study. The first interviews that I conducted were the most informal with people who work with tomato plants. Later, as I turned my focus to the bonsai and sequoia case studies, I codified my question list to better suit the normative issues that had emerged in these early interviews. These were informal interviews, however, so sometimes our conversations strayed from the questions detailed in advance. Each set of interview questions over different contexts was roughly the same, with some slight variations. Each interview included a request for suggestions of additional people to talk with about their relationships with a particular plant. For reference, you can see the list of questions included in the bonsai practitioner interview questions:

Bonsai Questions

Practitioner History and Context

1 How long have you been doing bonsai?
2 Do you work individually or with an institution?
3 What was your process of learning to practice bonsai?

Attitudes

1 What value do you think bonsai plants have?
2 What makes a bonsai good?
3 Is bonsai an art?
4 Which fits your attitude towards bonsai? Love, Use, Respect, Awe, Intrigue, Indifference, Scientific curiosity, etc. …

Relationship

1 Do you feel anything working with bonsai plants?
2 Do your bonsai plants depend on you? Do you depend on your bonsai plants? How? Why?
3 Do your bonsai plants communicate with you? If not, how can you tell? If yes, how do they communicate?

Consequences

1 Have you had bonsai plants die because of choices or actions you took? What was that experience like
2 Has a plant you once cared for died in the care of another?
3 Have the plants under your care gotten sick or had an adverse reaction to your actions that surprised you? How did you remedy it?

Practical Questions

1 How do you get most of your bonsai plants? (hunting, cuttings, seed, buying starts, etc.)
2 Who are your favorite bonsai practitioners?
3 Which type of plants are your favorite to work with? Why?
4 Why do you like bonsai?

Final Questions

1 Is there anything else that you think I should know that I haven't asked about?
2 Can you think of anyone who would be good to talk with further about this project?
3 If I have additional questions, can I follow up with you via email?

The questions provided cover a range of topics related to that person's relationship with bonsai that might be relevant to care ethics. If you recall from Chapter 3, vital elements of a plant-inclusive care ethic include valuing-attitudes from the carer to the cared-for, affect from the carer towards the cared-for, dependency within the relationship between carer and cared-for, responsive interactivity from the cared-for to the carer, and success conditions of care. Some of these crucial elements were missing from many how-to books and memoirs about plant carers. Consequently, the interviews were able to directly fill in these blanks. I asked questions about the carer's attitudes towards bonsai plants (e.g., What value do you think bonsai plants have? What makes a bonsai good? Is bonsai an art? Which fits your attitude towards bonsai: Love, Use, Respect, Awe, Intrigue, Indifference, Scientific curiosity, etc. ?). In these questions, I aimed to be direct but not leading. In some cases, I would follow up to clarify. In the question "Which fits your attitude towards bonsai?" I provided a list of possible attitudes, hoping not to lead too directly to a prescribed or hoped-for answer. The questions under the relationship section (e.g., Do you feel anything working with bonsai plants? Do your bonsai plants depend on you? Do you depend on your bonsai plants? How? Why? Do your bonsai plants communicate with you? If not, how can you tell? If yes, how do they communicate?) aim to understand if the essential qualities of an ethically caring relationship (e.g., affect, dependency, and responsive interactivity) are present. And the questions under the consequences section (e.g., Have you had bonsai plants die because of choices or actions you took? What was that experience like? Has a plant you once cared for died in the care of another? Have the plants under your care gotten sick or had an adverse reaction to your actions that surprised you? How did you remedy it?) aim to clarify the success conditions of the caring relationship from the view of the carer. I also ask general questions about their history with bonsai and practical aspects of their work with bonsai. This information helps me round out parts of the case studies that may relate to the ethical components of the relationships but are not as central to the question of ethical care being present in these relationships.

Some variations are present when considering the list of questions provided for the people who work with sequoia trees. In general, the sequoia tree interviews occurred further into my research from the tomato plant and bonsai tree ones. Consequently, some questions were changed to appropriately respond to friction or lack of fit in previous interviews.

Sequoia Questions

Practitioner History and Context

1 What is your personal or family history with plants?
2 How long have you been aware of Sequoia trees? How long have you interacted with Sequoias?
3 Do you interact with them as an individual or as part of an institution?
4 What was your process of learning about and how to care for sequoia trees?

Attitudes

1 What value do you think sequoia trees have?
2 How do other humans around/with you react to sequoia trees?
3 Which fits your attitude towards sequoia trees? Love, Use, Respect, Awe, Intrigue, Indifference, Scientific curiosity, etc. …

Relationship

1 Do you feel anything working with/being around/interacting with sequoia trees?
2 Do the sequoia trees depend on you? Do you depend on the sequoia trees? How? Why?
3 Do sequoia trees communicate with you? If not, how can you tell? If yes, how do they communicate?
4 Do you consider sequoia trees part of your community? If so, why? If not, why not?
5 Does your work with sequoia trees translate into your identity? If so, how? If not, why not?

Consequences

1 Have you experienced the destruction or death of a sequoia tree? What was that experience like?
2 Have the trees under your care gotten sick or had an adverse reaction to your actions that surprised you? How did you remedy it?

Practical Questions

1 Which type of plants are your favorite to work with? Why?
2 Are there trees that you value more than sequoia trees? If so, which one(s)? If so, why?
3 How do you think humans can best provide care for wild trees?

Final Questions

1 Is there anything else that you think I should know that I haven't asked about? Which questions should I have asked?
2 Can you think of anyone who would be good to talk with further about this project?
3 If I have additional questions, can I follow up with you via email or phone?

As you can see from the question lists provided above, in the bonsai interviews I asked if practitioners considered bonsai to be art. This is a subtle way at probing whether the interviewee thought that the bonsai has intrinsic aesthetic value as a work of art, or intrinsic moral value as a form of life or being. However, bonsai is a practice that occurs within an artistic and aesthetic context. Sequoia trees are more familiar in a wild context. So, I did not include the question about art for the sequoia practitioners. Additionally, there were some things that appeared more relevant for the sequoia context than the other two contexts. For example, "Do you consider sequoia trees to be a part of your community?" may have been more relevant to Native, indigenous, or First Nations practitioners.

The sampling was self-selecting (i.e., people who responded to my email or phone call) and snowball (as provided in the question sets above, I reached out to people who were suggested by other interviewees to talk to about their relationships). For a publicly accessible example of an interview, you can access my discussion with Ryan Neil on an episode of the *Asymmetry Podcast*, produced by Bonsai Mirai (Asymmetry Podcast 2022).[3] I took notes and recorded the interviews to refer to while writing.[4] Some interviews were conducted via telephone and could not be recorded, so notes were taken. Additionally, some interviews were able to be done in-person and on site. These included the Rainbow Tomatoes Garden and Chase Rosade's bonsai studio. The tomato plant interviews were far fewer than the others and were geographically close to me at the time of the project. The bonsai tree interviews were wider on a geographic scale, but still primarily within the United States. This is likely due to the snowball sampling and also my skillfulness in English, and less skillfulness in Japanese. The sequoia tree interviews were the most friction-filled, both in timeline and approaching interviewees. I believe this was the case because of my relative geographic distance (I was on the East Coast of the United States at the time and giant sequoia trees grow in a very limited area within the western part of the United States).

As Taylor notes, the creation of a comprehensive NCS methodology is not yet complete. As more philosophers engage in NCS, more examples are available to provide content for developing appropriate guardrails and construction guides. Taylor offers normative credibility as one part of a meaningful evaluation. "I propose that the soundness of NCS depends on its normative trustworthiness. … With normative trustworthiness, I refer to the degree of confidence that the case study offers a strong foundation for normative inquiry" (Taylor 2024, 309). I aim through these case studies to demonstrate that the relationships between human carers and plant cared-fors offer such a rich territory for inquiry. After arguing outside of an applied context for the inclusion of plant cared-fors into a care ethic, these case study chapters aim to demonstrate the promise and peril of applying such a theory to real life contexts.

The Point

The aim of this project is not to tell the reader what, empirically, all, most, or a particular percentage of the plant caretakers working in these contexts think about their work with plants. Rather, by applying the theory developed in the earlier chapters, I aim to examine whether the theory conforms with practices on the ground. Through dialectical engagement (e.g., asking and listening to plant carers about how they understand their work, their plants, their relationships), I hope to better understand the experiences of contemporary plant care takers and the caring relationship, replete with values and feelings; things not often explicitly included in plant care manuals. According to these plant care takers, is there anything more than a barely-qualifying-as-relationship with a (for all intents and purposes) green machine that happens to be living? Or do these relationships have value and emotional contours, a character that suggests more than just the hypothetical parts required for ethical care. Not merely could an ethically caring relationship exist in a thought experiment, but is the soil rich with such ethically caring relationships in the past and moving through the present into the future? While care ethics as a subdiscipline of philosophy is relatively "new," the components of ethical care can be found in numerous philosophical and historical traditions. And so, in the pages that follow, I begin sketching out the lines that historically build the pieces of our existing relationships with these plants. Out of these broad strokes emerge a more detailed account of a particular collection of plant caretakers in relationship with plants at a very particular time and place. From there, in conversation with these care takers, I seek to better understand the places of friction with the theory (e.g., where the theory developed in part one fails to account for the reality on the ground) and areas of fit. Additionally, there may be places where there is not friction, but space. In this space, I hope to identify future areas of research to better understand where we should go from here to understand and instantiate care for plants. In this way, the work of these case studies is dialogical. I have ideas that I enter into the interviews with—informed by prior research and personal experience—and the conversations I have with caretakers inform my understanding and revise my questions. Through the interviews, there are different threads that emerge to be more important to the ethical contours of the context then I knew before the interaction.

And so, these normative case studies that follow incorporate the work of historians, botanists, farmers, crafts people, home gardeners, foresters, and artists to create a picture of how care can be instantiated in three distinct contexts. Each context reveals a set of strengths for adopting a care ethic in the relationship between human carer and plant cared-for. Importantly, each also reveals certain areas of difficulty for the completion or achievement of good or successful ethical care. As with much of applied ethics, the devil can be in the details. Interdisciplinary work is painstaking. Boundaries can be drawn, transgressed, re-drawn, constructed, renovated, redesigned, or razed. Thankfully, I have predecessors who have developed conceptual tools and language for describing and evaluating this work. I am indebted to those who have embarked on these multi-tendrilled projects before. I aim to be clear with my methodology and trajectory as I navigate this complex ground.

Notes

1 Some argue that animal ethics (in our case, care ethics involving other animals) is distinct from environmental ethics. There is a history in the literature as seeing animal ethics as a step along the road to environmental ethics (for example, Aldo Leopold's evolutionary view of moral considerability in his *A Sand County Almanac*) or as a competing ethic entirely (Leopold 1949).
2 There is some debate over whether critical plant studies and the plant humanities refer to the same field. The former sometimes emphasizes a slightly different theoretical bend, while the plant humanities are less united by a particular theory.
3 Neil, a prominent U.S. based bonsai practitioner, requested that we record the interview as an episode of his podcast. Half of the episode is an interview with him in the style of interviews used for this project. The other half is his interview with me about this project. Available here: https://bonsaimirai.podbean.com/e/kate-brelje/.
4 Transcripts were not created nor are they available. This is one area of my approach that may depart between normative case studies and qualitative case studies.

References

Alarcon, Margaux and Pascal Marty. (2024). "Observing Farm Plots to Increase Attentiveness and Cooperation with Nature: A Case Study in Belgium." *Agriculture and Human Values*, 41 (2), 525–539.

Anthony, Raymond. "Building a Sustainable Future for Animal Agriculture: An Environmental Virtue Ethic of Care Approach within the Philosophy of Technology." *Journal of Agricultural and Environmental Ethics*, 25(2) (2012), 123–144.

Cañada, J. A. "Blocks and Overflows of Care: Territorialization and Marine Conservation in the Balearic Sea." *Science, Technology, & Human Values*, 0(0) (2024).

Davis, Dána-Ain and Christa Craven. *Feminist Ethnography: Thinking through Methodologies, Challenges, and Possibilities.* Bloomsbury Publishing, 2016.

Gagliano, Monica. *Thus Spoke the Plant: A Remarkable Journey of Groundbreaking Scientific Discoveries and Personal Encounters with Plants.* North Atlantic Books, 2018.

Gibson, Prudence. *The Plant Thieves: Secrets of the Herbarium.* New South, 2024.

Gilligan, Carol. *In a Different Voice: Psychological Theory and Women's Development.* Harvard University Press, 1982.

Giraud, Estève. "Urban Food Autonomy: The Flourishing of an Ethics of Care for Sustainability." *Humanities*, 10(1) (2021), 48.

Goralnik, Lissy, Matt Ferkany, Laurie Thorp, and Kyle Powys Whyte. "Philosophy in the Field: Care Ethics, Participatory Virtues, and Sustainability." *Resilience: A Journal of the Environmental Humanities*, 1(3) (2014), 1–28.

Hourdequin, Marion. *Environmental Ethics: From Theory to Practice.* Bloomsbury Publishing, 2015.

Knobe, Joshua and Shaun Nichols. "Experimental Philosophy." In Edward N.Zalta (ed.), *The Stanford Encyclopedia of Philosophy*, Winter 2017 Edition. https://plato.stanford.edu/archives/win2017/entries/experimental-philosophy/

Krzywoszynska, Anna. "What Farmers Know: Experiential Knowledge and Care in Vine Growing." *Sociologia Ruralis*, 56(2) (2016): 289–310.

Leopold, Aldo. *A Sand County Almanac.* Oxford University Press, 1949.

Neil, Ryan. "Kate Brelje." Asymmetry Podcast. Bonsai Mirai, 2022. https://bonsaimirai.podbean.com/e/kate-brelje.

Parsley, Kathryn M. "Plant Awareness Disparity: A Case for Renaming Plant Blindness." *Plants People Planet*, 2(6) (2020).

Seymour, M. and S. Connelly. "Regenerative Agriculture and a More-Than-Human Ethic of Care: A Relational Approach to Understanding Transformation." *Agricultural and Human Values*, 40 (2023), 231–244.

Taylor, Rebecca M. "Methodological Reflections on Normative Case Studies: What They Are and Why We Need Better Quality Criteria to Inform Their Use." *Educational Theory*, 74(3) (2024), 302–311.

Thompson, Merisa S. "Alternative Visions of "Ethical" Dairying: Changing Entanglements with Calves, Cows and Care." *Agricultural and Human Values*, 40 (2022), 693–707.

Wohlleben, Peter. *The Hidden Life of Trees: What They Feel, How They Communicate—Discoveries from a Secret World* (translated by Jane Billinghurst). Greystone Books, 2015.

5 Beyond Bondage

The Ethics of Caring for Bonsai Trees in Aesthetic Contexts

Bonsai as an Aesthetic Context of Care

Those who venture outside the mall in Washington, D.C. will be impressed by the carefully cultivated museum that is the National Bonsai and Penjing Museum. A place of living memories; plants marking time with their training, with the changing of their ownership, with their diplomatic and cultural meanings, with their changing bodies. One of the grandest trees, in my opinion, is the Yamaki Pine. In training since the early 1600s, the tree survived the atomic bombing of the Japanese city of Hiroshima by the United States government in 1945 (McClellan 2016, 34; Siddiqui 2015). In 1975, as diplomatic relations between the nations of Japan and the United States were recovering from the deep and enduring scars of war, this tree arrived in the U.S. As the National Bonsai and Penjing Museum notes, the caretakers were not even aware of how old the bonsai tree was until 2001 when the grandsons of Masaru Yamaki visited the living museum described above and told the story of the tree (McClellan 2016, 33): it is a tree surviving a nuclear holocaust; surviving a war effort that made keeping it alive even more difficult with scant resources; surviving the journey from Japan to the U.S.; surviving until this day, and hopefully on into the future.

This small anecdote indicates the historical and cultural significance of such trees. They are useful to nations and practitioners, for conveying deep meaning and participating in a long-rooted tradition of aesthetic practice. But as living things, one might also pause to consider the ethical dimensions of bonsai. What is the Hiroshima tree, morally speaking? And is the practice that has maintained its life over the past 400 years right or good? These questions will be addressed throughout this chapter, through the lens of care ethics, as a case study applying the normative ethical theory developed in the proceeding chapters to the specific case of contemporary bonsai practice.

While many of the plants in our lives we encounter within an aesthetic context, bonsai has certain peculiarities that prompt ethical concern. We are familiar with gardens with non-edible annuals, shrubs, and trees ornament public spaces and private dwellings, flowers as gifts that communicate gratitude, appreciation, or a job well done, and decorative houseplants are offered

DOI: 10.4324/9781003603665-6

as housewarming presents and host gifts. Plants provide a pop of color to a drab interior and improve air quality in office spaces and homes. These aesthetic contexts provide spaces for common human-plant interaction. Sometimes plants are offered by carers for other humans as an act of care (e.g., when a plant is offered as a gift for someone convalescing from an illness). But in this case study, bonsai is an exaggerated aesthetic case. While gardens could be a popular choice for such an examination (Cooper 2006), the testimony of bonsai practitioners and the unique set of relationship factors that bonsai requires (e.g., high dependency, generational care practices, and claims of pseudo-kinship) make it a good candidate for an in-depth examination of care ethics.

Bonsai is the practice of growing trees in trays that developed from the Chinese practice of penjing. Penjing was brought to Japan along with other cultural content by Buddhist monks around 1150 CE (Singer and Valavanis 2012, 12). But the popular practice of growing miniaturize trees in trays or pots is most readily understood as bonsai in the contemporary American context. Bonsai is a unique aesthetic practice. Historically and contemporarily, bonsai can be characterized in a number of ways. In the linage of Buddhist monks traveling from China to Japan, bonsai could be understood as a religious practice (Kobayashi 2018, 15–16). Though, even within this religious practice, there are deeply aesthetic elements. In a less religiously and philosophically rich context, bonsai can be understood as decoration. Bonsai is typically displayed in a special part of the Japanese household, the tokonoma, when it is at the height of its seasonal beauty (18). The rest of the time, bonsai plants are typically housed outdoors. Some understand this as an invitation to meditate, while others consider it a pretty decoration that enhances the living space. Bonsai can also be seen as an art. Many contemporary practitioners that I talked with pointed out the fact that bonsai is a living art form.[1]

The primary overarching puzzle in this case study involves both the fit between the plant-inclusive care ethic developed in the first half of the project and the relationship between human carer and plant cared-for in the context of bonsai. It is also an exploration of particular ethical issues that emerge from this relationship and practice. Is it possible and practical for the carer to have valuing-attitudes for the bonsai plant? Does the relationship between the carer and bonsai plant include affect, dependency, and responsive interactivity? Can the practice of bonsai ever produce success conditions that contribute to the thriving of the plant? And finally, what other ethical concerns emerge from dialogical engagement with the practice and testimony of practitioners? Throughout this chapter, I find two particularly relevant areas that emerged in my discussions with practitioners: valuing-attitudes and success conditions. Claims of kinship and actions that reflect the value of the plants used in bonsai (e.g., setting up a care regimen in one's last will and testament for continued care of the plants) contrast with claims about bonsai plants being standard art objects. Also, the largely bifurcated responses to bonsai plants (either of attraction or disgust) suggest a need to interrogate whether bonsai as a practice could ever actually meet the requirements within care ethics; primarily, the question of whether the bonsai plant can flourish.

Beginning with valuing-attitudes, does this process of artistic engagement reduce the bonsai plant into a mere object for manipulation? If the bonsai process requires a type of objectification of the plant such that the plant is only valuable as a tool for the artist's expression, this may appear to be a clear violation of care ethics described in the previous chapters. The pure instrumentalization of a plant undermines the valuing-attitude required by the carer for the cared-for. Yet, many people who work with bonsai appear to have valuing-attitudes towards the plants that are more in alignment with an ethically caring relationship. When I was young, I remember having a strong impression that took root in my brain that bonsai artists were in communication with the plants, and that the structures and forms that the plants took were a type of coordination between human artist and plant. With that view of the human-plant relationship, the plant would resist objectification through being granted some agency within its co-creation of form. This case study aims to interrogate the practice of bonsai to understand how, when, and if these conflicts emerge.

When bonsai is displayed human responses seem to bifurcate between fascination and repulsion. This reflects the concern about the success conditions required for good care within a care ethic. Are bonsai pampered and well cared for plants, or starved and tortured trees? Is the training of miniature trees a practice of active co-creation that is compatible with ethical care? Or is it an example of domination, where the human agent paternalistically and oppressively determines the plant's shape and size, restricting growth to serve a wholly human end?

There are other aspects of the relationship between human carers and bonsai cared-fors that are well suited for a care ethics case study. The dependency of the plant on the human carer is evident in the name of the practice. Bonsai plants are by definition planted in containers (by strict definition, trays). They tend to be smaller than they would grow otherwise because their space for growth is limited by the container. These limits are imposed by humans in bonsai. Some trees naturally stay small and are limited in growth potential by natural geographic and ecological factors (e.g., a tree that grows in a rock crag). But bonsai requires intervention from a human caretaker. The bonsai practitioner prunes the tree, fertilizes it, shapes it, and sometimes changes its container. Bonsai trees, properly maintained, can live for centuries.

Another reason that I focus on bonsai is because the bonsai literature in English is rich with care language (e.g., Yoshimura and Halford 1957). So, practitioners refer to the relationship between practitioners and the plant as one of care. Bonsai plants are also typically highly valued by the practitioners and the institutions they are affiliated with. Gardens, arboretums, and individuals invest large resources into the maintenance, health, and wellbeing of their plants.

I focus on contemporary American bonsai practice in this case study. As I'm writing this analysis from the U.S., these case studies examine plants that I have a biographical connection with and also demonstrate care in varied contexts. Also, interestingly, American intervention has played a role at several points in the history and current practice of bonsai. As I explain below, some think the exposure of bonsai to military personnel after WWII led to its proliferation abroad. Bonsai is also a distinct aesthetic tradition and has vast well-articulated conceptional resources in English; in written word, video demonstrations, and oral tradition. Additionally, bonsai seems to be an affective and moral flashpoint for humans. Some human viewers are aesthetically repulsed, others aesthetically attracted. Importantly, some are morally repulsed, and others are morally attracted to the practice. Bonsai is rich with aesthetic and moral judgments from practitioners and audience members alike which provides a good context for making sense of ethical care. Finally, my grandfather was a bonsai practitioner in the U.S. for a time. He was a hobbyist for a few years when I was young. I fondly remember my grandfather taking me to his bonsai club meetings and purchasing a beautiful ficus for me. Now both the tree and my grandfather are dearly departed, but I think it is important to apply our ethical theories in our own lives.

Brief Bonsai History and Introduction to the Process

The Chinese practice of *penjing*, a form of landscape-based art in pots using rocks and plants, predates bonsai (Singer and Valavanis 2012, 12). As mentioned before, it is thought that Buddhist monks brought *penjing* to Japan, where it became bonsai. Bonsai occurred throughout Japan, but in one documented case it was a unifying element of a town. In 1925, Omiya Bonsai Village was established as a small settlement of bonsai practitioners outside of Tokyo (Suga 2019, 79). The practitioners who relocated there from Tokyo developed the town around bonsai, providing yards for the potted plants. During World War II, bonsai was generally considered unpatriotic (80). Water and labor were supposed to be diverted from pre-war pursuits to support the war movement. As will become clear in the testimony of contemporary practitioners, bonsai is a time and resource intensive practice. The residents of bonsai village would most likely have spent considerable time working with their plants. Many bonsai plants need to be watered every day. This requires constant labor, attention, and water resources that would detract from a nation-wide shift to focus nearly exclusively on the war effort. While a few practitioners, like Kyuzo Murata, defied governmental and cultural pressure to give up their trees during the course of World War II, many were not able to avoid the draft (80–1). Allied bombing campaigns (nuclear and otherwise) also undoubtedly killed a number of bonsai trees. Overall wartime was unkind to the Japanese practice of bonsai.

In a strange twist, the U.S. presence in the aftermath of World War II was a boon to the international bonsai community (80–1). Omiya Bonsai Village managed to survive the war. American and international service personnel stationed in Tokyo could visit Omiya Bonsai Village, which likely encouraged American fascination with bonsai and helped it spread internationally post-war. Books written in English about bonsai were published in the 1950s. Two important examples of such works are Tatsuo Ishimoto's *The Art of Growing Miniature Trees, Plants and Landscapes: Japanese Bonsai and Bonkei adapted to Western Conditions* (1956) and Yuji Yoshimura and Giovanna M. Halford's *The Japanese art of miniature trees and landscapes: their creation, care, and enjoyment* (1957).

The Bonsai Process

Creating a bonsai tree starts with pre-bonsai plants. These plants are produced in one of three ways: starting from seed, cuttings, or hunting. Starting a bonsai from seed, while common in novelty gifts like The Mini Bonsai Kit (2001), is a difficult process for individual bonsai practitioners. It requires a lot of patience, as seed-started trees may need to grow for eight to ten years before starting the bonsai shaping process (Rosade 2022). A more accessible version of propagation for bonsai practitioners is growing trees from existing plants. A tree is pruned, and a cutting from that pruning can be propagated into a new plant. This can be done through placing the cutting in a container with water to promote root growth or applying root hormones, depending on species and growing conditions. To be clear, these need not be cuttings only from other bonsai trees. In theory, any small tree cutting could hypothetically be a pre-bonsai plant. Another propagation method is through layering, where a part of the existing plant develops roots before separating the section from the mother plant. Many practitioners obtain pre-bonsai plants from nurseries. In Japan, different nurseries care for trees for different growth periods (Strehse, Troyer, and Mahler 2022). In the U.S., it is less structured by tree age. Practitioners can buy from local nurseries who source their bonsai plants from other specialty growers.[2] Some practitioners and studio owners grow their own starts from cuttings or seed (Rosade 2022).

A third option for obtaining a pre-bonsai plant is hunting. Trees collected from the wild are called *yamadori*, which roughly translates from Japanese to English as "plants from the mountains" ("Yamadori" 2022). *Yamadori* pre-bonsai are highly valued, as they can be centuries old and have a great amount of character and aesthetic quality born out of growth related to the natural environment. Dan Wiederrecht and Steve Varland from Backcountry Bonsai call these trees "survivors" (Wiederrecht and Varland 2022). If you were out for a hike, *yamadori* pre-bonsai plants would be the small trees that appear like they're barely making it, like the infamous Charlie Brown Christmas tree. Despite appearances, these trees can be healthy, in the sense that they're still alive and growing after dozens or hundreds of years in rough natural conditions.

Some practitioners hunt on their own, but given the manifold legal, ethical, and biological issues involved with hunting, it is important to carefully consider the expertise of the hunter. Practitioners who go out on their own without knowledge of root systems, collection techniques, and bonsai practices may very well maim or kill a number of otherwise healthy trees living their lives in the wild (Wiederrecht and Varland 2022). At least one interviewee also flagged collection as a key hotspot for ethical concern (Leskoviansky 2022). The interviewee did not claim that pre-bonsai plants should never be collected from the wild, but rather said that the practice of collecting is fraught with ethical dimensions in a way that starting plants from seeds or cuttings is not. This is in part because the plant has been living, in some cases for hundreds of years, in the wild. When one collects a plant for bonsai, they are transferring the plant to a state of near-total dependence on human caretakers. This results in the tree-collector (and then owner) taking on a great responsibility, because the tree otherwise could have continued living just fine, and is now in a state of need from a human caretaker. There are also legal issues and corresponding philosophical questions of ownership and wild plant use that emerge in *yamadori* collection.

After a pre-bonsai plant is selected and reaches a state of good health, the process of training the plant into a bonsai can begin. The pre-bonsai is planted in a tray (or a shallow vessel) that could be a variety of sizes or shapes. As a tree ages, it will be repotted over time. Sometimes this means a tree will grow and require a larger tray or pot, which can in turn encourage more growth. When a tree is re-potted, the roots are usually trimmed and specialty blends of soil are used for better root development that is conducive to maintaining the health of a tree in a miniaturized state (Hagedorn 2022). Another common horticultural procedure used in bonsai is pruning branches.

A key practice in bonsai is tree shaping. The relatively small size of the plant and the use of shaping practices are two of the most recognizable characterizations of bonsai. Bonsai trees are shaped primarily through pruning and wiring. As Emily Brady and Isis Brooks explain in "Topiary: Ethics and Aesthetics," pruning, and even styling trees, does not hurt them. Rejecting sentience due to a lack of a nervous system, pruning itself does not get in the way of a tree's goal-orientation (Brook and Brady 2013, 129). So, the tree itself is unharmed unless the styling makes it impossible for the tree to thrive. A common method is by moving branches and wiring them into position. Some tree branches are moved to the point that they creak under the pressure of the formation. Of course, forcing a tree into an untenable position could damage the tree or branch that one is seeking to manipulate. So, bonsai practitioners must use horticultural knowledge to not push trees "past the point of no return." Improper training (either by pruning or shaping) could endanger the life of the tree. But successful bonsai practitioners minimize the trees they kill and maximize the trees that grow and maintain their health under their care.[3]

Training bonsai trees is a complex practice. Some American practitioners create a naturalistic style (Strehse, Troyer and Mahler 2022). This style seeks to highlight natural characteristics that trees might have in the wild. Mentioned briefly above, *yamadori* trees can be plants that already have attained a certain character that bonsai practitioners look for in a tree. This might be age, asymmetry, or an interesting trunk shape. These trees grew in this way because of their natural conditions; a rock crag, shallow soil, wind on a cliff face, heavy snow accumulation, a windstorm, etc. When bonsai practitioners acquire these trees, they do interact manipulatively with them. They change more about the tree: they limit tree growth by planting it in a tray, ground trees in the trays, wire branches, trim branches, etc. Practitioners also do this with trees started from seed or cuttings.

One major indication of inappropriate wiring is when it cuts into the tree. This is seen as bad within the practice of bonsai (meaning, it misses the mark aesthetically and indicates a lack of skill when manipulating the tree). There may also be aesthetic judgments because the cords make the tree look unnatural, destroying the illusion of naturally looking miniature tree. From an ethical perspective, the wrongness of this may be from it harming the tree in some way, if the wiring impairs the organism's health. It could also undermine the tree's autonomous contribution to its shaping, ignoring the natural way of the tree; something a practitioner should work with, not against.

Bonsai trees are traditionally kept outdoors in their trays. Exceptions include tropical plants, or plants that are housed in inappropriate (or non-natural) climates. It is typical for practitioners to have an outdoor cold frame or other housing for plants over winter, so that they are protected from falling branches, wind gusts, and other extreme weather events. This allows them to still go through the typical seasonal cycles without risking too much harm (or unintended sculpting) from destructive elements. Occasionally bonsai trees are brought indoors to display during a time that they are blooming or looking particularly radiant. After the display, the tree is moved outdoors again. It is during display time that wires and supports might be removed as well. After the display, these implements can be reapplied.

Contemporary Practices

There are three primary groups that I discuss as contexts for bonsai practitioners: hobbyists; professional practitioners, teachers, and artists; and curators. Each category provides a distinct context for practice that highlights different areas of valuing-attitudes, caring relationships, and success conditions.

Hobbyists

Hobbyists are amateur bonsai practitioners. As one interviewee suggests by designating beginner hobbyists from advanced hobbyists (Bello 2021), the spectrum of experience and expertise amongst hobbyists is vast. Hobbyists can include someone who just got their first plant and attended a workshop, or someone who has practiced for a few decades and regularly volunteers with the bonsai collection at their local arboretum. Corresponding details of particular hobbyists' attitudes towards plants, recognition of caring relationships with plants, and adherence to success conditions for the care of the plants are as varied as the practitioners themselves.

One example of a hobbyist is my grandfather, Joe Driver (1931–2022). He was a member of his local bonsai society, a nonprofit group for hobbyists with the goal of education and community building (Strehse, Troyer and Mahler 2022). He had his own collection of bonsai trees. When I was young, I accompanied him to one of his bonsai club meetings. He bought me a tree—a beautiful ficus. I was not properly trained, and the idea of pruning or wiring the tree was distasteful to my mother. So, instead of growing it like a bonsai, we took care of it like a houseplant as ficuses don't thrive outdoors through all four of our seasons. We were lucky that ficus trees can survive under that treatment for longer than trees that need to be left outside (though protected and cared for in particular ways) during the winter. Eventually, the beautiful ficus succumbed to our poor care after living with our family for several years (including traveling with me for my final year of undergraduate and to Colorado for my master's degree). I admit now that I did not do right by that tree. I painstakingly helped it with two bouts of insect infestation (one that resulted in my small apartment smelling heavily of fish oil for over a month from applying a general "organic" pesticide spray, and another that required wiping each leaf down with soap and water). Like other unwitting bonsai plant caretakers, I did love that tree, but my skills and horticultural knowledge (and likely the lack of prioritization of learning proper techniques) were not enough to keep it alive.

I am not sure what became of my grandfather's bonsai plants. My aunt said that he left them to die after he was no longer able to care for them. This is not necessarily the practice of all bonsai hobbyists, though it does seem to suggest something about my grandfather's attitude towards the plants. In many other cases, collectors and practitioners will designate caretakers in the event of their passing. Some with large collections even donate them to local arboretums where they can receive care. Professionals definitely have a different dedication for their plants and keep in mind the need for a plan to assure care after the practitioner's death. Some hobbyists may feel this need to provide a care plan for their plants after their passing, but it depends on the individual.

For this project, I talked with the board of the Pennsylvania Bonsai Society, a during-the-time-of-first-writing local-to-me nonprofit bonsai club. The Pennsylvania Bonsai Society was formed in 1963 (The Pennsylvania Bonsai Society). Three board members, Bill Strehse, Jim Troyer, and Robert Mahler were able to talk with me. Strehse and Troyer are advanced hobbyists, while Mahler is a professional practitioner. Strehse started bonsai as a young person, then took a long hiatus before returning to practice for the last six years. Troyer has been a long-time practicing hobbyist. Mahler was fascinated by bonsai at a young age, eventually going on to apprentice in Japan for 4.5 years and curating the bonsai collection at the Brooklyn Botanical Garden upon his return (Mahler 2022). After curating, he established Kifu Bonsai in East Greenville, PA.

The goal of the bonsai society is manifold, but primarily they want to build human communities, expand interest in bonsai, and share educational resources about bonsai. The Pennsylvania Bonsai Society hosts traveling teachers, holds workshops, and displays annually in the Philadelphia Flower Show (PFS). In working with the public, the board members note that audience responses to the bonsai trees at the PFS tend to be positive, e.g., looks of wonder, beauty, and amazement (Strehse, Troyer and Mahler 2022). While some audience members, including Mahler's own grandmother, have negative responses to bonsai, these tend to reflect the concern over "torturing" the trees. This will be addressed in length below as I interrogate the possibility of good outcomes for trees in the bonsai context.

Professional Practitioners, Teachers, and Artists

Similar to hobbyists, each professional practitioner, teacher, and artist in the U.S. context has a distinct approach and set of underlying beliefs about bonsai trees. Some of these practitioners consider the work they do with the trees as art. Others define bonsai as a craft. For some professional practitioners, they consider themselves primarily teachers. Within the contemporary bonsai context, it is usual for people to start as hobbyists that attend presentations, workshops, and join local bonsai groups. Then, as they become more interested in bonsai, they may study with a "bonsai master," traditionally in Japan, though not all bonsai professionals take this path.

After training for a sufficient amount of time, advanced bonsai hobbyists may transition to the professional sphere. This just means that they devote their livelihoods to bonsai. That often entails having their own bonsai studio, where they offer classes, board client plants when the owners are traveling, and keep their own plants. They have areas to store plants during winter and inclement weather, as well as display areas for the spring, summer, and fall.

There are several bonsai practitioners in the southeast region of Pennsylvania that were interviewed for this project. Jim Doyle of Harrisburg, PA, is a teacher and owner of Nature's Way Bonsai in Harrisburg, PA. He took classes from Chase Rosade, a bonsai practitioner, teacher, and studio owner of Rosade Bonsai based in New Hope, PA. The three board members of the Pennsylvania Bonsai Society also attended workshops with Rosade. He used to travel extensively to teach bonsai techniques and engage with the international bonsai community. While Doyle and Rosade are bonsai artists in their own right, their current focus is on education.

Another educator and bonsai practitioner interviewed for this project is Ryan Neil, who founded Bonsai Mirai, a community based in Portland, Oregon. Interested in bonsai since age 12, Neil pursued an education in horticulture. He completed a five-year[4] apprenticeship with Masahiko Kimura, a leading (if not the top) bonsai practitioner in Japan today. Neil founded Bonsai Mirai when he returned to the States. He does work that utilizes Japanese techniques in a more American naturalistic style and is considered a top U.S.-based bonsai practitioner.

Curators

For this project, I spoke with three curators of bonsai collections: Andy Bello, a museum specialist at the National Bonsai and Penjing Museum in the United States Arboretum in Washington D.C.; Carmen Leskoviansky, curator of the University of Michigan bonsai collection at the Matthei Botanical Gardens and Nichols Arboretum in Ann Arbor, Michigan; and Arthur Joura, curator of the bonsai collection at the North Carolina Arboretum in Asheville, North Carolina.

Curators provide care for plants that are owned by public and private institutions. These curators may have their own private bonsai practice aside from their curatorial duties. I interviewed the curators like the rest of the bonsai practitioners, but it is important to note that their work requires a slightly different orientation than hobbyists or professional practitioners. Unlike hobbyists, bonsai is part of their livelihood while they have their curatorial position. Unlike professional practitioners, they are less reliant on the sales of their trees or success of their classes to make ends meet. Curators, like nonprofit bonsai associations, are very concerned about education and outreach. Increasing engagement with bonsai collections is a priority, alongside cultivating informed viewership. At least two of the curators that I interviewed began working with bonsai having more horticultural expertise than bonsai experience.

Though each curator has to uphold some tradition in the styling of the trees in their care, they also bring their own vision to the work they do at museums and arboretums. Bello mentioned a special interest in creating bonsai that reflect particular places in the United States, as well as connecting with indigenous plant experts to understand bonsai from Native American perspectives. Leskoviansky is interested in exploring gender and bonsai, and expanding the possibilities of bonsai. Joura creates landscapes that resonate with North Carolinians, echoing ecosystems throughout his state. It is an exciting time to observe what bonsai curators are doing across the United States.

Application and Analysis

To test my theory of plant-inclusive care, I must first address whether bonsai practices can cohere with three crucial ingredients of ethical care:

1 valuing-attitudes from the carer towards the cared-for,
2 a caring relationship between the carer and cared-for that has caring affect, dependency, and responsive interactivity, and
3 consequences of caring actions for the cared-for are beneficial for it.

Valuing-attitudes

Per my argument for ethical care, the carer must have attitudes that ascribe intrinsic moral value to the cared-for (in this case, the bonsai plant). As mentioned in Chapter 3, part of why I leave this open to the attitudes ascribing value to the cared-for is because of the great breadth of scientific, religious, and philosophical paradigms that underly particular plant care takers' relationships with plants. There is a huge spectrum of views on inherent moral value as it relates to plants, and my goal in this book is not to defend one source of value against another. Rather, I want to understand if ethical care is possible in the human-plant relationship. As will become clear here, the breadth of attitudes that bonsai practitioners hold for their plants identify a wide array of potential sources of intrinsic moral value.

The most common attitudes reported by practitioners were the valuing-attitudes of reverence and respect. The source of these attitudes towards the bonsai plant depended on the individual. One practitioner stated that "Plants are people too" (Leskoviansky 2022). Another interviewee made interesting assessments about different valuations of particular trees. They said that after working with particular trees every day they become an "individual" (Joura 2022). At that point, for that interviewee, the bonsai tree is set apart from others, after all the work and the relationship that they have built and experienced with that "individual" tree. When given the option of choosing from a pre-selected group of possible attitudes (love, use, respect, awe, intrigue, indifference, scientific curiosity), a few practitioners chose "all of the above." One practitioner chose "all" until they re-read the list and saw "indifference" and corrected that they hold all of the positive ones. Most interviewees emphasized having love for their bonsai, which was followed up with describing their bonsai plants as "like family members" (Joura 2022, Wiederrecht and Varland 2022). I think it is reasonable to infer that a family-like love assumes a type of intrinsic moral value in the plant.

When I asked practitioners what the value of the bonsai is, the responses were wide-ranging. A few practitioners identified bonsai as valuable because it is "the only living art form" (Doyle 2021). In some sense, the value of a bonsai tree is derived from the fact that it is living, so it may have something to gain or lose from its continued existence. Alternatively, it could be because it is unique in the use of living entities for artistic creation (Doyle 2021). The value also might come from the audience reaction to the tree that it creates an emotional response, similar to definitions of successful artistic execution. In these cases, aesthetic and ethical values both blur together and occasionally emerge distinct. If a bonsai is valuable because it is an artistic product or creation, then that is an aesthetic value that might be distinct from any moral value. So, if the bonsai is valuable because it elicits emotions from an audience or viewer, then the judgment of value is more based on aesthetic value (e.g., beauty, sublimity, ugliness, horror) than moral value (e.g., instrumental or inherent). Similarly, if the valuing-attitude is one of appreciation as a living art form, unless there is some specific source of moral value alongside the aesthetic value it has as a work of art, the valuing-attitude simply suggests that it is an aesthetic being, not necessarily an ethical one.

This is an especially important point to make in the context of bonsai; particularly because it is considered an aesthetic practice. Care ethics demands that the cared-for is not a mere object (even if it is valuable as an art object). The cared-for does not qualify as a participant in a caring relationship because it is beautiful, ugly, or sublime. These are aesthetic qualities and have corresponding judgments of aesthetic character or value. While ethics and aesthetics can (and frequently do) intersect in fascinating ways, being clear about the claim of value being made by the bonsai practitioner is key for understanding whether an ethics of care is properly applied in this context.

I think one indication that could indicate more than an aesthetic judgment and more of a valuing-attitude of the practitioner is the sourcing for caretakers for the trees when the practitioner is unable to care for the plants. One way this happens is when individual collectors or educational or nonprofit institutions invest in plant specialists who can take care of the plants. This suggests to me that the plant is worth training and employing one or more humans for the full-time care of the plants. Many institutions began their bonsai collections when a previous owner gifted them at a point when the collector could no longer take care of the plants. When bonsai practitioners can no longer care for their plants, many make plans for their care. Some start selling off or redistributing plants to knowledgeable caretakers, while others who are still in a position to take care of their plants prepare for their plants to be transferred at the time of the practitioner's death. Thus, the trees are valuable beyond the life of the carer.

However, again, it is important to distinguish whether this is done because the plant is considered to have aesthetic or moral value. For example, one might find a new caretaker for a painting or textile when one is no longer able (financially or physically) to keep the artwork in a safe environment. This is done perhaps because the owner or caretaker believes it has aesthetic value that is worth protecting. In the case of aesthetic art objects, it can also be a reflection of concern for the instrumental (market) value of the work. Rather than valuing the painting or textile as a purely aesthetic affair, the owner views protecting the work in the same way they might as defending a financial investment. This, I think, is a crucial indicator of valuing-attitudes that attribute inherent value (moral or aesthetic) to the plant. And the clarity around what the ethical theory developed in previous chapters is clear: the carer must hold valuing-attitudes (moral ones) towards the plant cared-for. I think in some cases this reflects the reality of valuing-attitudes held by practitioners (as in the case where bonsai plants are considered persons), but other times practitioners fail to hold non-aesthetic valuing-attitudes for the plants under their care.

What this means in practice, on the ground, is that to better adhere to a care ethic, the carer in the relationship with a bonsai cared-for should develop valuing-attitudes towards the plant that are not merely aesthetic. The bonsai plant cared-for can certainly be valued in part because of its aesthetic qualities, but to adhere to the theory I propose, it must also be valued morally. Whether this is because the bonsai practitioner believes the bonsai plant is a person, or because it has intrinsic moral value as a living being, or in virtue of its ability to flourish or because it is kin, the valuing-attitudes of the carer for the cared-for create the conditions of an ethically caring relationship.

Caring Affect

Interviewees' discussions of their relationships with bonsai were rich with affect associated with care. While in some interviews I asked directly about affect, many interviewees peppered our conversations with affective language unprompted. They expressed positive feelings for the plants, such as "love" and "joy" (Leskoviansky 2022). And when things were not going well for the plant, one practitioner reported "feeling sorry for the plants" (Joura 2022). For this practitioner, the bonsai they ended up working with were passed down from a previous collector who could no longer take care of them. The interviewee felt "miserable because of what was being done to them" (Joura 2022). When a tree dies, practitioners reported feeling badly. For one, the experience was described as "heartbreaking" (Wiederrecht and Varland 2022).

Some of these emotions and feelings (e.g., love and joy) are explicitly connected to care ethics (Noddings 2013). They reflect an investment in and response to the cared-for's flourishing. However, if you recall from above, a claim that some aestheticians will make about the value of an art work or aesthetic experience is that it evokes a certain kind of affect. The mere presence of affect may not be enough to confirm that the relationship is a caring one.

The feelings reported on the death of a bonsai plant are important for teasing out possible moral dimensions of care in the relationship between carer and cared-for. This is a controversial claim, but it is important to explore. When a bonsai plant dies, people tend to feel something. It could be a feeling that one has on behalf or for the plant who has passed. Other times, it might be a feeling of disappointment of the work that one has put into the plant. Some others might feel distress at losing the financial investment made in the care and maintenance of the plant. But I think it is something more than this. Of course, humans can feel many emotions and feelings simultaneously. We are complex creatures. The negative affect that we feel when a plant under our care dies can be all about us (e.g., a shame felt for failing to keep the plant alive, or a disappointment in the loss of an artistic object we appreciated). It can also be a response to a perceived loss for an other, namely for the plant who has passed. There could be a sadness that the plant is no longer living and experiencing the world it once had, or an anger that the plant no longer has a future for itself. But there is also the possibility that the affect is reflective of a relational loss. Particularly in the care required by a bonsai plant, a carer can feel as though there has been a loss in their life of a companion, friend, or family member. Bonsai practitioners spend a lot of time and energy with their plants. As we'll find below, at times the bonsai plant is the center of the carer's world. The loss of that relationship can prompt affect just as easily as its presence.

That is why I think in applying a care ethic to the human-plant relationship, an openness to and interrogation of affect is so important. I think that "negative" or weighty emotions can be as revealing as pleasant ones. What are we mourning when a plant dies? Is it the reflection on us as a caretaker? Is it the life of the plant for the plant's sake? Is it the loss of our relationship with the plant? Any of these are possible, and I think it is particularly telling when we look at the emotions around the death or loss of plants what the contours of our relationship with plants are (or could be). We will revisit this in Chapter 7, where the death of giant sequoia trees looms large in the affect connected to the caring relationship.

Dependency

The relational dependency of the bonsai practitioner and bonsai plant seems straightforward from the side of the plant. In terms of physical dependency, humans are constantly intervening on the bonsai plant and its environment to keep it alive and healthy. Because the plant is in a hyper-vulnerable position (living within a tray), the bonsai plant is wholly dependent on the human carer for water, access to the right amount of light, proper soil, up-potting when necessary, and protection from elements and pests that might threaten the plant's life.

Human practitioners are often dependent on the bonsai plants in physical and social ways. For professional practitioners and curators, caring for bonsai plants is their livelihood. Beyond that, practitioners report developing better character traits and becoming better people because of their bonsai practice. Others report beneficial impacts on their mental health. One practitioner noted that their students reported the importance of bonsai for "getting them through the (2020 COVID) pandemic." And as noted above, when a plant is lost, a practitioner can feel as though a family member has died. For one bonsai practitioner, bonsai gives their life meaning. In these myriad ways, human practitioners depend on their bonsai plants.

Responsive Interactivity

Bonsai practitioners report constant interactivity with the plants. When asked if the plant communicated with them, the answer was a resounding "yes." One respondent humorously retorted that how a plant was communicating with them "depends (on) how much you've been drinking" (Wiederrecht and Varland 2022). All laughing aside, the practitioners emphasized how the plant constantly responds to their actions. Humans also constantly respond to the activities of the plant. Some of the language used by practitioners about communication was, in their own words, anthropomorphic. Practitioners discussed the necessity of listening to the trees (Neill 2022); both "listening to the movements" and "hearing the fibers crack/creak" (Bello 2021). In some cases, this was described as visual interpretation, e.g., "how to read the tree's life experience" (Wiederrecht and Varland 2022). So, both audio and visual elements came up as ways to communicate with trees.

Attentiveness seems essential for the successful care of bonsai plants. Many serious practitioners will be constantly connected to weather information and, when need be, will wake up in the middle of the night to move or protect the trees under their care (Bello 2021). Practitioners may need to have assistance in moving or protecting trees (because of tree size or practitioner ability.) Nearly every practitioner interacts with each tree every day. The most common interaction is through watering. This constant attentiveness to the tree aids in catching health issues early which can be essential in maintaining a tree's health.

The human-plant relationship is central in the practice of bonsai. One teacher actually begins their classes and workshops with meditation and emphasizes the individual human connection with a particular plant (Doyle 2022). At least two practitioners identified bonsai as a "lifestyle." One practitioner admitted to waking up in the morning thinking about bonsai and going to bed thinking about bonsai after spending the whole day working on bonsai (Rosade 2022). Of course, many of these practitioners are professionals or curators, but hobbyists are also required to interact with their bonsai plants on a daily basis to ensure the plant is healthy and respond to any discoloration, insect infestation, or change in shape that is indicative of the plant's faltering health.

A few interviewees discussed "entering into a partnership" with the bonsai plant (Bello 2021, Neill 2022). In their view, bonsai is not a mere manipulation of a plant that totally bends to the will of the human practitioner. Rather, the practitioner must be able to "listen" to the plant, through observation and interaction. In some cases, this is described as a "cooperative union" (Joura 2022) where the practitioner is "entering into a relationship with" (Neill 2022) a tree, and bonsai work something the practitioner is "doing with the tree" (Strehse, Troyer and Mahler 2022).

Other times, practitioners consider the bonsai plant to direct the relationship. One practitioner stressed the importance of "follow(ing) the tree's lead" (Strehse, Troyer and Mahler 2022). Another likened bonsai to another aesthetic activity, wherein the skilled practitioner is "letting the tree guide the dance" (Neill 2022). The tree has some basic shapes and attributes. If the practitioner attempts to manipulate the tree in a way that goes beyond the organism's physical limits, then parts of the tree can break or wither. As one practitioner put it, the "'natural movements' (of the) tree leads design" (Bello 2021). The partnership requires buy-in from both the plant and the practitioner.

Bonsai practitioners claim that if a plant is not happy, it will not do well. "A healthy tree is a happy tree" (Strehse, Troyer and Mahler 2022). Happiness, usually a distinctly human or animal capacity, is probably a stand-in for doing well, being healthy, or flourishing. If the actions of the practitioner undermine that welfare, health, or flourishing, then the plant falters (e.g., gets ill, succumbs to infestation, loses parts of its body, etc.). So, the practitioner approaches the plant with careful observation and a commitment to support the health of the plant overall, and the plant does its best to first survive the interventions, and then thrive in the new shape and environment. In this way, the practitioner is "work(ing) with" a being that can and does interact with your actions rather than working on an inert material (Doyle 2022).

One way to understand this relationship is that the practitioner is put in the position as a learner from the plant or a student of the practice of bonsai. One interviewee explained that in bonsai they are "being taught" (Bello 2021) while another considers themself a "student of bonsai" (Rosade 2022) rather than a master. One practitioner was adamant that "Bonsai plants teach me more than I teach them" (Doyle 2022). Practitioners, even professionals and teachers, consider themselves to be constantly learning about and from their plants.

Beneficial Consequences of Caring Action for the Plant

To address the concern over ethical care, we have to ask what successful care of a bonsai plant looks like for the plant. Some ways of understanding what is good for the plant might be longevity in the life of the plant, being a healthy plant, successfully reproducing or continuing a genetic lineage, or flourishing in whatever way that looks like for that particular type of plant. All of these are controversial options in bonsai, and even if non-controversial in theory, are difficult to pin down what it looks like in practice.

As one practitioner notes, all living things die (Bello 2021). This is why longevity might be more important to the plant than total avoidance of death. All practitioners admitted that plants (plural) have died under their care. As they become more proficient in their care, fewer die on their watch. Tree life-spans, bonsai or otherwise, are the culmination of myriad factors. The environment, genetics, and intervention of animal activities (human animals included) can all influence individual plants and their prospects for longevity. Some practitioners mention that trees under the care of a set of good bonsai practitioners (because ideally, a bonsai will go through the hands of a few generations of caretakers) have a greater likelihood of surviving longer than in nature (Doyle 2022). Bonsai trees under successful care can live for a long time. For example, according to one interviewee, there were some bonsai trees in Japan when they trained there that are around 1,100 years old (Strehse, Troyer and Mahler 2022).

Another possible answer of what is important for the plant itself is overall health. If I keep my aloe plant alive, but it's constantly teetering on the edge of dying, I have a hard time imagining that longevity is all that might be important for that plant. Health, a somewhat conceptually murky term when used within the human context, might best be understood in this context as the state of a biological organism with a good amount of nutrients, a robust immune or defense system, and a good environment which allows the organism to live well. Health in a bonsai is evaluated through horticultural tools. Bonsai practitioners regularly and consistently research methods for how to improve their care of the plants. As several interviewees admitted, the first step in practicing bonsai is establishing a healthy plant. Successful stylization and manipulation cannot be accomplished without the plant being healthy first.

A third possible issue of concern to a plant might be reproduction. This outcome might assume a type of aim or goal that the plant has. While valuation of one's genetic code may be an invention of humans projecting purposes or goals to living beings where they may otherwise not exist, the behavior of organisms that reproduce can suggest some value in reproduction. We may be able to look to the lives of other living beings to understand if reproduction is the determining factor in a life well lived. Certainly, for humans, this may not be the case, as many of us do not reproduce our genetic material in the form of a biological child and still claim to live well. I imagine my dog's life is similar, that though he has not sired any puppies (that we know of), he could still be understood as living a good life. There might be other ends, or goals, than procreation.

Flourishing is an additional factor that might be important for the plant. Flourishing might be understood as identical to health, but I think in this case, it is important to separate them. This is because a plant could be perfectly healthy in a tray, but some might judge the tree to be prohibited from aiming towards some type of idealized potential. Being "allowed" or placed in a suitable context for "natural" growth could be seen as more conducive to flourishing than the dependent, growth-stymied intervention by humans that is required in the practice of bonsai.

One concern within flourishing is the relationship that the plant can have with other members of their ecosystem. If it is meaningful for the plant to connect with and respond to other organisms and environmental forces, similar to how we talk about animal welfare (both human and more-than-human), then making sure that the plant has an environment that supports (rather than inhibits) novel interactions and behaviors might be important. This is a place where bonsai may struggle in large part because *yamadori* plants are taken from an ecosystem and isolated in a pot or tray. The plant can still interact with other organisms in the ecosystem, but it might be much reduced in the breadth of organisms it is able to interact with compared to its pre-bonsai context. Because the plants roots are also cut, the plant is far more constricted in its reliance on humans to intervene in the management and facilitation of relationships the plant forms with other organisms and how it can respond. This may reduce or limit the range of expressions available to the plant, and if they matter to the plant, may impact its quality of life.

Well cared for bonsai plants can easily meet the first three of these conditions for beneficial consequences resulting from a carer's actions. Bonsai plants can live as long, if not longer, than what they could "in nature" (i.e., in a state prior to the intervention of a human digging them up and putting them in a pot). Bonsai plants can also be incredibly healthy and have the prioritization of their health as a (if not the) primary goal of their carers. And bonsai practitioners can facilitate reproduction by encouraging pollination, taking cuttings of the trees, and harvesting seeds or fruit. It is the last condition that gives pause when responding to the claim that practitioners are harming the plants in their care by undermining their flourishing.

On this point, in some cases, I think such a claim might be:

a slipping from an aesthetic judgment to a moral one,
b conflating the natural with the good,
c unfamiliar with the horticultural requirements of bonsai, or
d working within a philosophical paradigm that disapproves of a degree of human intervention in non-human lives.

For the first concern, a person might look at a bonsai tree and make an aesthetic judgment about the tree's appearance and slip into a moral judgment about the practice that created that tree. For example, a person interacting with a bonsai tree might judge the plant to be strange, disturbing, curious, or novel. Instead of keeping this judgement as a purely aesthetic one, it might lead into a further moral judgement about the intentional human-directed process that creates such miniaturized trees. In effect, something ugly could be simultaneously judged as morally bad. However, we know that in many cases, what is ugly, odd, or strange is not necessarily morally bad. Aesthetic judgments and moral judgments can coincide or relate, but they needn't. They can completely come apart.

A second way to interpret this judgment about bonsai practice could be conflating the natural with the good. Part of the concern over bonsai practice limiting the flourishing of the tree, is the stunted potential that the tree could hypothetically otherwise reach if taken out of the tray and planted in the ground. Part of this might overlook the natural state of some *yamadori* pre-bonsai trees, which occur naturally small in the wild. Miniature trees exist in nature, their growth inhibited by their natural environment and surroundings. But assuming this was not the case, the miniaturization of the tree could seem unnatural. What is unnatural is bad, in a similar way to what is natural is good. There are plenty of examples of natural occurrences that could be considered morally bad. It also puts forward a type of naturalism that must be comprehensively argued for. Needless to say, just because something is natural does not mean it is good, and just because something is unnatural does not mean it is bad. Furthermore, just because something is natural does not mean it supports or guarantees flourishing. And something unnatural could either support or undermine flourishing. The state of flourishing can be divorced from whether it is reached naturally or unnaturally.

One common negative response to bonsai by gardeners and non-growers alike is that the practice is a type of torture or starvation of the trees. If an organism is tortured or starved, that likely undermines the organism's flourishing. However, bonsai practitioners do not starve their trees. Bonsai trees receive water nearly every day to ensure their health. They also receive fertilizer. While their growth might be limited by the tray and the space available for their roots to grow, they are not starved. Starvation would undermine the health of the tree, which makes the bonsai practice impossible. Bonsai relies on healthy trees. In response to the claim of torture, one would first have to admit that plants are sentient. If plants can have sensations, then it is possible that these sensations might be provoked in negative, torturous ways. Such torture would likely be evident physically in the tree, through wounds, scarring, etc. Those markers are avoided in good bonsai practice. Some might say that not all torture results in physical scarring, but can be internal, psychological pain. Again, the person claiming that the plant has been tortured would need to explain sensations and a psychological life as a precursor to such torture. Perhaps this mistreatment would be evident through plant behavior, i.e., the way a plant responds to its environment. Again, one would first have to get on-board with the idea of plant behavior. If one admits plant behavior exists, then we would need to look for signs of plant distress. However, bonsai practitioners are keen observers to the "happiness" of their plants, and claim to strive to ensure a healthy plant. I am not sure how miniaturization alone would harm the plant itself.

Finally, one might judge human intervention with a plant as being at odds with the plant's flourishing. This view admits a negative quality to some human interventions with plants. These judgments can occur on a spectrum: from views that espouse any intervention with plant life is immoral to a context-dependent appraisal of intervention within a plant's life. In the case of bonsai, the human intervention with the plant is very intensive. The human is constantly intervening with the plant; e.g., providing water, moving the plant, trimming the plant, re-potting the plant, displaying the plant, treating the plant for infestations and illness, etc. Such interventions (including, perhaps most importantly, the confinement of plant to tray) might be understood as immoral. On this concern, one must show how or why this intervention is immoral, and how or why it goes against the plant's flourishing.

In determining the importance of success conditions for the community of bonsai practitioners, it is worth mentioning that the bulk of books in English about bonsai are guides for care. They emphasize particular techniques that are beneficial for human practitioners maintaining the health of the tree within the bonsai practice. As mentioned before, many practitioners value their bonsai plants inherently. The plant matters to the practitioner. So, having successful, beneficial consequences for the plants is important for bonsai practitioners. Sometimes these consequences coincide with the practitioner's artistic or manipulative aims of creating a tree in a particular shape. Bonsai texts provide detailed how-tos that include basic plant health care. Some recent texts, like Michael Hagedorn's *Bonsai Heretic* (2020) go a bit further to criticize and correct misconceptions about plant care for older bonsai trees. His heresy could be understood in some ways as bucking the traditional practice for the benefit of the plant.

Conclusion

In conclusion, some love bonsai trees, some hate them. Some view it as a respectful relationship with a plant while others see it as undermining a plant's flourishing. Does the interaction between bonsai practitioners and their bonsai plants cohere within a plant inclusive care ethic? I argue that yes, it does. Though bonsai is a horticultural practice that humans have strong feelings about, this analysis shows that while some might hold these beliefs, the bonsai context of practitioner-plant relationship can cohere with an ethic of care. Bonsai practitioners can hold valuing-attitudes that ascribe intrinsic moral value to their plants, have caring relationships rich with caring affect, dependency, and responsive interactivity, and have beneficial outcomes of caring action for the plants themselves. The bonsai process itself does not undermine the possibility of success conditions, despite concerns that these plants are starved or tortured. Rather, the plants can survive and thrive for hundreds of years if properly cared for by skillful carers.

This is not to say that all human practitioner relationships with bonsai plants are always morally good in real life. Bonsai can go awry at any point in the analysis of ethical care: some bonsai practitioners may fail to have the requisite valuing attitudes for the plant, others may engage with the bonsai in a relationship that is uncaring and inattentive, while still others can simply fail in their actions being beneficial for the plants. Assessments of bonsai from the general public seem to bifurcate between delight and disgust, praise and scorn. This judgment responds to the third criteria of care: the success or failure of caring actions. If one believes that bonsai is a practice that starves a plant or impedes the plant's ideal flourishing, then the judgment follows that the actions of care fail in all cases of bonsai. However, if one believes that bonsai is a practice that feeds a plant and encourages the plant's flourishing, then the judgment should follow that care has been successful. Ultimately, I argue that bonsai can be (and often times in fact is) a successful ethically caring practice. But the devil is in the details—certain conditions must be met, not only in the valuing-attitudes of carers, but also in the contours of the relationships they form with the plant cared-fors, and the resulting consequences of the actions for the plants.

Notes

1 Some interviewees described it as the only living form of art, but that is a highly contestable claim.
2 For example, I called my local nursery (Primex in Glenside, PA) and asked where they sourced their bonsai trees. They identified Meehan's Miniatures in Maryland—a regional specialist grower.
3 I am not sure if overall the life success rate or longevity of bonsai plants is better or worse than in nature. That might be something to consider.
4 Neil was actually working with Kimura in Japan for six years. The apprenticeship only lasted five. It is tradition to stay an extra year to "give back" to the community that educated you.

References

"About Us." *The Pennsylvania Bonsai Society.* www.pabonsai.org/about-us (accessed August 24, 2022)
Bello, Andy. Interview. December 2021.
Brook, Isis and Emily Brady. "Topiary: Ethics and Aesthetics." *Ethics and the Environment*, 8(1) (2003), 126–142.
Cooper, David. *A Philosophy of Gardens.* Oxford University Press, 2006.
Doyle, Jim. Interview. December 2021.
Hagedorn, Michael. *Bonsai Heresy: 56 Myths Exposed by Using Science and Tradition.* Stone Lantern Publishing, 2020.
Hagedorn, Michael. Interview. February 2022.
Ishimoto, Tatsuo. *The Art of Growing Miniature Trees, Plants and Landscapes: Japanese Bonsai and Bonkei adapted to Western Conditions.* W. Heffer and Sons, 1956.
Jeker, François. *Esthétique et bonsaï: guide pratique.* Letter. 2000.
Joura, Arthur. Interview. January 2022.
Kobayashi, Kunio. *Bonsai.* PIE International, 2018.
Leskoviansky, Carmen. Interview. January 2022.
Mahler, Bob. "Meet Bob Mahler | Japanese-Trained Bonsai Artist | Kifu Bonsai." *Kifu Bonsai.* www.kifubonsai.com/about-the-artist (accessed August 24, 2022).
McClellan, Ann. *Bonsai and Penjing: Ambassadors of Peace and Beauty.* Tuttle Publishing, 2016.
The Mini Bonsai Kit. Running Press Minis, 2001.
Neil, Ryan. "Kate Brelje." *Asymmetry Podcast.* Bønsai Mirai. 2022. https://bonsaimirai.podbean.com/e/kate-brelje
Noddings, Nel. *Caring: A Relational Approach to Ethics & Moral Education*, 2nd edn. University of California Press, 2013.
Rosade, Chase. Interview. 2022.
Siddiqui, Faiz. "This 390-year-old bonsai tree survived an atomic bomb, and no one knew until 2001." *The Washington Post.* August 2, 2015.
Singer, Jonathan M. and William N. Valavanis. *Fine Bonsai: Art and Nature.* Abbeville Press, 2012.
Strehse, Bill, Jim Troyer, and Robert Mahler. Interview. January 2022.
Suga, Y. 2019. "Branded heterotopia: Omiya Bonsai Village in Japan from 1925 to the Present Day." *Studies in the History of Gardens & Designed Landscapes*, 39(1), 77–89.
Wiederrecht, Dan and Steve Varland. Interview. 2022.
"Yamadori - GLOSSARY | BONSAI." https://bonsai.shikoku-np.co.jp/en/word/2009/04/post-179.html (accessedAugust 24, 2022).
Yoshimura, Yuji and Giovanna Halford. *The Japanese Art of Miniature Trees and Landscapes.* Tuttle Publishers, 1957.

6 Tangling with Taste

The Ethics of Caring for Tomato Plants in Crop Contexts

Caring for Crops: Care in the Context of Consumption

That tangy, pungent smell covers my arms after I emerge from my plunge into the bushy, overgrown spot behind my house. Five more red, small delicate fruits are added to my makeshift container for gathering this morning. Some split from the pressure of being pulled from the plant, others have half-healed scars from growing beyond the means of their skin. These tomato plants have taken over. Originally started from seed and subsequently planted in old plastic nursery pots, their roots must have extended beyond that barrier into the ground. There is no other way to explain their height (far exceeding my 5' something), vigor (spreading over other pots, rich with fruit), and expanse (growing into the windows at the back of the house). My hope is that they will re-seed themselves for years to come, leaving a presence and reminder for future renters that this is their home.

But is it? Tomato plants originate from the Andes mountains. While these two strains, Coyote and Matt's Wild Cherry, are more similar to the shape and style of those early wild tomato plants, western Virginia is not their native home. Here, they are not perennials but annuals; having to cycle through birth and death each year. I love tomatoes and seeing these plants grow gives me so much joy. Ultimately, though, I have chosen these plants because I like eating tomatoes. What is the point of my buying seeds from my local store to grow *Solanum lycopersicum*, if not in hopes that those seeds will yield delicious fruit for me to consume? This chapter examines human-tomato plant relationships to ascertain how care ethics can apply in a context of use and consumption. Something that seems to violate care in a morally praiseworthy parent-child relationship (for example, the parent using the child merely as a means to an end) is an action that is often observed in relationships between humans and crop plants. But ethics of care does not simply take what "is" and turns it into an "ought." Rather, it prescribes ethical attitudes and actions. So, in this chapter, we must examine the human-plant relationship that includes consumption to determine if ethical care in these most instrumental-appearing relationships is possible. As Wilson and Gillespie write,

DOI: 10.4324/9781003603665-7

> The question is some form of "how do we care about the tomato?" Or if that sounds too precious, "how do we care about and care for the world, even as we work it, depend upon it, need to eat it and walk over it?"
> (Wilson and Gillespie 1999, 208)

The plant-inclusive care ethic proposed in Chapter 3 highlights valuing-attitudes, caring relationships, and success conditions. The carer must have certain attitudes towards the recipient of care; these attitudes could be love, attentiveness, care, or some combination of these and others. At least one attitude must attribute some sort of intrinsic or inherent moral value to the cared-for. This signifies that the cared-for is in the realm of ethical concern and demonstrates that the carer judges them to have moral standing.[1] The caring relationship between carer and cared-for must also contain caring affect, dependency, and responsive interactivity. Finally, the success of the actions and practice of the carer must be evident in the life of the cared-for.

As will become apparent throughout this chapter, ethics of care might actually yield more fascinating possibilities for ethically right relationships between humans and crop plants. While the history of human-crop relationships provides both purely instrumental and richer ends-oriented treatment of the plants, in ethics we should not merely look to the world as it is and project that it is how it should be. Yet, in environmental ethics, knowing how things are on-the-ground, how things *actually* are in the world, provides fertile ground for criticism and expansion of our ethical theories that we seek to use in the real world. This background is important for understanding the relational context in which human-tomato plant connections occur. I will provide a brief history of the tomato plant in human societies, its path from wild to domestic, from the Andes to Italy, from decoration to nutrition. The history will lead to an analysis of contemporary tomato cultivators: industrial farms, home gardens, and seed saving collectives. This examination will show the richness of contemporary human-tomato plant relationships and the possibility of ethical care in them.

Caring attitudes combined with tangible success conditions allow us to speak meaningfully of care ethics including plants, even plants that human beings raise for use and consumption. Ultimately, I argue that ethics of care can provide a satisfactory account of care between human agents and tomato plants. Within the context of consumption, two components are the most challenging to meet: proper valuing-attitudes about the cared-for and success conditions. Regardless of scale, these can be an issue for farmers, gardeners, and seed savers. If practitioners address these concerns, then their relationship with tomato plants can cohere with an ethics of care framework even in a use and consumption context.

The Historical Context of Human-Tomato Plant Relationships

Human societies have considered tomato plants to be both threatening and nutritious, decorative and curative. Some tomato plants produce fruits keenly tied to national identity and family culture. The first known origin of the tomato plant is in the Andes region of South America (Smith 2001, 15). There is little archeological evidence of the pathways of the wild tomato plant into North America (Smith 2001, 15). What is evident, is that the precursor to the tomato plant that we recognize today was domesticated (or cultivated) in pre-Columbian Mexico (Smith 2001, 15). At that time, it was known as tamatl, tomatl, or xitomatl (Smith 2001, 15; Wilson and Gillespie 1999, 188–210). Spanish conquistadors brought tomatoes back to Europe and spread them throughout the continent.

Cookbooks and garden notes provide some evidence for the presence of tomato plants in a given cultural context. Tomatoes were listed in Bernardino de Sahagún's 1580 *General History of the Things of New Spain* (DeWitt 2015, 27–8). Tomato fruit began appearing in cookbooks in the late 17th century (DeWitt 2015, 170–1). By the early 18th century in Rome, Francesco Gaudentio, includes in his text *Levitico*, which discusses the food sources of Jesuits, that "tomatoes were 'cultivated in gardens'" (DeWitt 2015, 172). A little over a century later, Italian cookbooks would sometimes omit particular recipes for tomato sauce "because 'everyone knows how to make it,' whether from fresh, preserved, or sundried tomatoes" (DeWitt 2015, 175). Tomato plants had taken root in European homes and kitchens.

Within the North American context, it is unclear exactly how tomatoes arrived in the United States; but it was probably through Mexico, Jamaica, or Europe. Andrew Smith argues that linguistic evidence best supports the Jamaica to U.S. pipeline (Smith 2001, 20, 26). Whichever path brought them to this area, human-tomato relations went through massive shifts in a relatively short period of time. In the early 1800s, the fruit of tomato plants were generally considered toxic (Smith 2001, 3–4). The best use for tomato plants was as a decorative element in the garden, because of the plant's colorful, visually appealing fruit.

Eventually the tomato plant transitioned from mortal threat and disgust-inducing botanical to a source of medicine. Tomato plants were documented inhabitants of Thomas Jefferson's Garden at Monticello in the early 1800s (Monticello 2022) and listed in the 1807 catalogue for Bartram's Garden in Philadelphia, PA (Bartram 1807, 33). There are numerous American stories recounting "the first man to bravely eat a tomato, at his own peril, in public" (Smith 2001, 4–5). This heroic tale signifies the shift from poison to non-lethal edible. In the 1830s, tomatoes were advertised as miraculous medicines (DeWitt 2015, 180). Tomato pills were sold with the promise of curing everything from rheumatism to syphilis (Smith 2001, 108). Eventually, tomatoes were accepted in the U.S. and transitioned to a food source.

Within the long history of human-plant relationships, the cultivated (domestic) tomato plant is a fairly recent addition, but it has a rich context. There are other crops that have longer histories, some even tied to our species' very precious Neolithic agricultural revolution when we transitioned from hunter-gatherer societies to land-locked agrarian ones (Blakemore 2019). But I chose this plant because of its contemporary ubiquity in contemporary global cultures, its rich documentation of shifting attitudes throughout human communities, and its place in society as a crop. Tomatoes have also been a mainstay of our family garden and where better to begin to apply one's ethical theory than at home?

Cultivation: The Importance of Domestication

The tomato plants that most of us reading this text around the time of its publication are familiar with are cultivated. Wild tomato plants still exist in the Andes region, but they are not the tomato plants that are ubiquitous in the world today (Smith 2001, 15). Domestication is a crucial part of the tomato plant's story within human societies and also an important facet of care relationships—particularly as they relate to femininity. To help illuminate this, turning to the animal ethics discourse can be useful, as contemporary animal ethics deal with some of the same issues that plant ethicists may face and have already developed critical ethical frameworks for considering non-human entities.

The animal ethicist Clare Palmer (2010) utilizes context to formulate a relational ethic between humans and animals.[2] Noting morally relevant contexts (e.g., histories of interventions by humans in the lives of domestic animals, lack of intervention in the case of wild animals) she shows how and why there are different moral obligations to wild and domestic animals. This is important for the case studies approach I am taking. One large contextual difference between the tomato plant and sequoia tree case study is the differentiation between crops and wild organisms. Some wild plants exist in extractive or consumption-focused relationships with humans, but crops are constantly intervened upon and manipulated by humans. Humans go to great lengths to secure certain crop seeds, seedlings, planting areas, and the product of those plants: vegetables, fruit, and seed. They crossbreed different types of plants to enhance the goal of consumption (e.g., taste, hardiness, shape, health, size, etc., of the plant, fruit, or seed). And humans dramatically change the land, landscape, and ecosystem to support the cultivation of crops.

One defining characteristic of crops is the process of domestication, "the evolutionary process whereby a population of plants becomes accustomed to human provision and control" (Vodouhè and Dansi 2012, 1). Cultivation occurs when a specific plant is raised in a designated space intentionally by an agent. Often, the agents placing and tending to plants in a particular area do so in order to get something from the plants, like medicine, food, or bast fibers for textiles. The tomatoes we consume today are the product of selective breeding, happenstance, human choice, and human intervention over long periods of time.

Another implication of domesticity in connection with plant life is its association with femininity. Plants are one of the few areas of Anglo-European science that have a history of welcoming the participation of women of high class within some modern cultures. While some herbalists are associated with occult women's knowledge that could be considered dangerous by the dominant culture, Christian traditions within the Western world (e.g., witches, pagans, other non-Christians who are close to plant religious and healing traditions, Hall 2011), others were considered the appropriate use of even the good Christian women's time and energy (Norwood 1993). The tamed garden was an extension of the home. At times, this domestic claim extended beyond the garden and into the wilds of the North American continent, which were considered to also be a space of potential home for the settlers and colonial descendants. Of course, in the dreaming of establishing this home, settlers and colonists would both destroy the existing homes of native, indigenous, and First Nations peoples, and the ecosystems in which plants, animals, insects, and other living beings had already established their "homes."

I mention this for two reasons.

First, ethics of care is considered an ethical theory that in part emerges from feminist criticism of masculine theories. Care could be considered a feminine ethic, in that it uplifts values and relationships that are considered feminine (e.g., mother and child, dependency, etc.) Whether these are truly feminine values or relationships is another discussion altogether, but the ethical theory that I'm utilizing in this work is connected to the domesticity associated with human women in some dominant cultural narratives. Second, in the realm of science, the domesticity connection between women's roles and plant life allowed women to openly study botany, and occasionally participate in horticulture (Norwood 1993).

The inclusion of women in the sciences can be powerful. Depending on one's philosophical framework, this inclusion may yield better scientific results (see Longino 1990). Despite this promise of greater diversity of scientists yielding better science, there are glaring shortcomings of the claim that the inclusion of women was radical in diversifying botany in the 18th and 19th centuries. Many of the women involved were of a certain class and race. The domesticity associated with botany was a particular type of domesticity, in current parlance a type of white feminism pertaining to the European descendants taming and making a home out of a "New World." That is not to say that women of different classes and races were totally removed from the study and use of plants. But the coupling of botany and European domesticity at a specific time and place in history is one of a plurality of domestic and womanly ideals.

One particularly interesting case of women scientists working with crop plants is Barbara McClintock's work on maize, as described by Evelyn Fox Keller in *A Feeling for the Organism*. While McClintock did not readily identify herself as feminine, or her scientific methodology keenly tied to her gender or sex, Keller pulls out differences in her practice that show the importance of her inclusion as a historical outsider in her scientific sphere. McClintock famously made huge breakthroughs in transposition, at the time a little-known phenomenon in genetics. She worked steadfastly with her plants. Keller recalls her interview with McClintock in the final chapter of *A*

Feeling for the Organism: "No two plants are exactly alike. They're all different, and as a consequence, you have to know that difference," she explains.

> I start with the seedling, and I don't want to leave it. I don't feel I really know the story if I don't watch the plant all the way along. So I know every plant in the field. I know them intimately, and I find it a great pleasure to know them.
>
> (Keller 2003, 198)

Keller uses this passage to support her argument that good scientific inquiry should have a place for emotion and feeling. And while McClintock's aim is more scientific inquiry that happens to be housed within agriculture, her description of attentiveness is something that is shared with the testimony of gardeners and farmers. While the scale and ability to know each plant might be different, whether plant care-takers were working at an industrial farm, in their backyard garden, or in containers, close observation of the plants is central to care. These differences in context of plant rearing will be discussed more in depth below.

An ethicist of care might happily flag "feeling" as a sign that McClintock's sex or gender did indeed play a role in her scientific practice. The relationship between subject and object relayed in "feeling" highlights affect or a felt (real or imagined) closeness between the scientist and her maize. Is this feeling empathy? Familiarity? Insight? Regardless, feeling creates an affective tone that is compatible with the caring relationship that this chapter emphasizes. McClintock doesn't seem to ascribe personhood to her maize, and yet affect, attention, and care are present in her treatment and understanding of the maize other.

Briefly mentioned before, the more obscured and demonized (literally) plant knowledge and practice of women who do not appear in upper classes, or who are not members of a given race or ethnicity, or do not adhere to accepted religious traditions can put them in peril. The cultivation of plants for medicinal purposes not connected to masculine masteries over unruly nature has historically been dangerous for women in these positions (e.g., witches and healers), and continues to be (e.g., plant abortifacients). Despite the presence of plant cultivation and even domestication, the aesthetics of an evil witch living in the woods (see Eggers 2015) may lean towards wildness rather than domesticity. Despite the same practice of cultivation and care of the plants by the queer and unruly solitary witch elder creating a salve from foraged plants for arthritis as the barefoot and pregnant heterosexual, monogamous wife chopping herbs from her garden for her child's headache, the former is a warning if you stray from the path and the latter a beacon of state approved femininity.

All this to say, that domesticity in both care ethics and the growing context of plants must be examined with an awareness of cultural influences and layers of meaning. As will become evident in what follows, contemporary context of care for tomato plants can be largely industrial. While the family resemblances of things considered "feminine" and "masculine" vary, the role of aesthetic judgment in these contexts can unwittingly usurp the moral, as with the witch and housewife. Beyond the aesthetic differences of small-scale gardening and industrial farming, there is a looming question about care. At first blush, care might seem to fit small-scale farming best because of its aesthetic aura of idealized feminine domesticity. A woman laboring around her home to cultivate children and plants, goes to her thriving home garden to pick healthy and homegrown vegetables to nourish her family. While this might be a simple description of a woman doing an everyday activity, moral values could be projected onto it, especially if we are not mindful of the slippage between the aesthetics of care and the ethics of it.

We live in a time where a vision of a glorified, non-existent past is held up as a beacon for a return to the earth (and the heteronormative nuclear family, and patriarchal values, and religious fundamentalism). Care could be uplifted as a tool of domination and oppression in this political and ethical social movement if cut off from its feminist roots. Gender looms large in our current plant contexts, and as the bonsai chapter demonstrated, aesthetics can have a profound impact on our ethical assumptions and values. We should be mindful of this as we conjure images of 1,000-acre fields plowed with massive machinery and humble apartment patios speckled with pots. The contexts of cultivation, even when looking at a single type of plant (in this case, the tomato plant) are diverse and varied.

Contemporary Contexts for Growing Tomato Plants

For tomato plants, I want to focus on three particular contemporary contexts: farming, home gardening, and seed saving collectives.

Tomato Plants and Farming[3]

Within this project, I consider farming to be determined primarily by scale and method. The use of particular technologies on a different scope distinguishes farming and industrial agriculture from gardening and seed saving collectives. Farmers have goals that prioritize larger scale production for a consumer market. This requires the development of industrial infrastructure at the farming and processing levels of the tomato industry. Historically, there was an intentional process towards greater mechanization:

> In the first generation of harvesters, unripe tomatoes were spotted by people riding in rigs who picked them out and threw them off. The newer harvesters, such as those featured at country fairs and at the Woodland Tomato Festival, now "read" tomato color photoelectronically as the tomatoes come up the belt and punch unripe ones back into the field with a rank of little hydraulic plungers.
>
> (Wilson and Gillespie 1999, 205)

In this country fair scene, the harvest of fruits from tomato plants is organized around efficiency and effectiveness. That occasionally requires development of infrastructure that supports faster planting, rearing, and harvesting; be it mechanized or human labor. While mechanization is in greater demand, part of what farmers argue is driving this technological development is a labor shortage (Harper 2021). The amount of labor required (as an alternative to or complement of mechanization) depends on the type and use of the fruit.

This use-oriented design in farming also makes demands of the plant itself. As tomatoes grew in popularity, there was a reasonable increase in the demand. To produce and distribute large quantities of the fruit, some characteristics of *Solanum lycopersicum* needed to change. For example, G.C. "Jack" Hanna worked for years to crossbreed tomato plants in aims of creating a tougher tomato fruit that could withstand industrial handling. His success cemented his position in the agricultural industry as the "father of 'the square tomato'" (Wilson and Gillespie 1999, 200). The square tomato was a great industrial breakthrough because its shape was able to better withstand the rough treatment of industrial technology in reaping and moving the fruit.

Furmano's is one example of an industrial processor that relies on farms to produce large quantities of tomato fruit. While Furmano's is not as large as top U.S. tomato product processors, they are a local-to-me industrial processor who has welcomed contact and inquiry from me about their farms and industrial processes. Furmano's utilizes top industrial methods.

The growing process for Furmano's farms starts with seed acquisition. Furmano's and other industrial operations buy rather than save seed. This is in part because of the types of plants that they tend to grow hybrids rather than heirlooms (Weaver 2021).[4] Hybrids are plants that result from crossbreeding two different types or varieties. Often this pollination process is done manually, to ensure the correct purity of the seed produced. This is the case of staple tomato plant hybrids like Burpee Big Boy. Seed savers, who we'll discuss below, also sometimes create hybrids to sell or supply to smaller seed companies. The choice of hybrid or heirloom seed depends on the variables relevant to the gardener or farmer. In some ways, hybrid seed could be more expensive for a farmer because it costs more money to buy seed every year from a seed company. Hybrids tend to be more expensive than heirloom seed because they are more labor intensive to create and you cannot harvest the seeds from hybrid tomatoes and expect all of the seed to yield the same type of tomato plant as the one you planted from bought seed. Also, some companies and consumers think the existence of "hybrid vigor," a greater resistance to pests because their characteristics are the result of a good crossbreed, justifies a higher price for hybrid seed.

Furmano's procures their seeds, many in the Heinz variety, from California and Canada (Hoffman 2021). Then they germinate the seed in their greenhouses and cultivate seedlings. When the time is right, the seedlings are planted in the farmer's fields (for Furmano's, most of those fields are in Pennsylvania, New Jersey, and Maryland) using RJ Equipment planters that mechanically insert small plants (plugs grown in greenhouses) into the soil (Hoffman 2021). When the tomato fruit has ripened, harvest begins. Furmano's uses Pik Rite Tomato Harvesters, based in Lewisburg, PA. Pik Rite harvesters are machines that cut the tomato vines, collect tomatoes and plant parts on a conveyor belt, and sort fruit from other plant matter. Red tomatoes are moved through a chute into a large vehicle container that drives alongside the harvester as it tills up and sorts the plant parts. Green tomatoes are separated out from red and disposed of with the other leftover plant matter in the field. The plant matter and green tomatoes can be used for fertilizer. Furmano's practices vertical tilling, though they have experimented with no till practices. They follow industry standards of IPM (integrated pest management) and regularly undergoes sustainability audits (Hoffman 2021).

Ownership is an important facet of contemporary farming. Some farmers rent the land on which they grow plants. Some farmers do not own the seed or the plants themselves, but rather are like contractors to a larger processing corporation (similar to the chicken farming industry). Some farmers grow on smaller scales and look more like (or simply are) family farms that sell to producers who include different things in their contract with the purchasing company, e.g., providing seed, allowing farmers to use expensive equipment like planters, harvesters, etc. (Harper 2021).

Consumption is the ever-present concern of the farmer. How will this product sell on the market? What will the yield be? The scale of the market and the identity of the consumer (spanning continents, foodways, cultures, and geographic territories) determines what is grown, and how much of it is grown. In the case of industrial tomato production, if the tomato plants are the cared-fors, who are the carers? To be a proper carer in care ethics, one must be a moral agent, since ethical care is not merely a custom or arbitrary set of rules, but a practice that carries moral weight. So, considering the current conditions of robotics in farming, these machines fall short of being moral agents. They are unable to understand morality and partake in moral reasoning. So these mechanical harvesters would not be moral agents and could not be considered the ethical caretakers of the plants.

To find the morally responsible carers on the farm, we must find the moral agents. These could be the land owners who decide what happens on a given lot of ground, the humans involved in the seed industry determining which seed is sold and what qualities are desirable in a plant, the humans in the fertilization and pest control industries, and the humans on the ground running the planters and harvesters.

Some environmental ethicists no doubt consider some types of farming (i.e., industrial agriculture) a moral non-starter. In the animal world, industrial agriculture has been a perennial source of grave immorality for animal rights activists, animal ethicists, climate ethicists, and human rights ethicists (see People for the Ethical Treatment of Animals). One might assume that some of the same criticisms of animal industrial agriculture (e.g., confinement, hazardous waste production, water use and pollution, ecosystem disturbance, removal of humans from the process of rearing the things they consume) are all similarly present in the case of large-scale plant agriculture (see Rollin 2016, Cocksaw 2021, Lundqvist 1998, Li 2017, Peck 2010). One major difference may be sentience. Assuming that tomato plants are not sentient, the immoral realities of other animals existing in conditions of torturous suffering within many industrial agriculture operations are not present in the industrial procurement of plants.[5] Animals (including humans) may suffer (i.e., experience pain) from pesticides, pollution, and ecosystem services loss through the tendency of contemporary farming methods to wipe out local biodiversity. But the most damning ethical and aesthetically repulsive criticisms of industrial animal agriculture undoubtedly revolve around sentience and abuses of sentient creatures in virtue of their ability to suffer.

There is another way that industrial scale farming might be a concern for care ethics. As mentioned previously, if a plant has a stake in its ecosystem being rich with other forms of life, then perhaps certain agricultural techniques (e.g., planting monocultures, controlling non-plant life to produce the biggest yield of commodified plant products, etc.) would undermine the plant's ability to meaningfully engage with its environment. Inhibiting a plant's ability to respond to novel stimuli, or navigate ecosystems with more than a few heavily controlled organisms, might make life worse for the plant. If one were to grant this type of meaning for plants, then current large scale agricultural practices might be at the very least limiting a plant's flourishing, and more robustly, undermining the possibility of it.

Additionally, some might be concerned about other effects that industrial agriculture has on plants themselves—particularly with seed and growing practices (Curry 2022). Industrial agriculture can create conditions where biodiversity is lost because a large demand for inexpensive produce may limit what is produced on an industrial scale. As is clear in the history of the tomato in industrial agriculture as the plant needed to be changed to accommodate machinery. So, there are families of plants that could be difficult to process on an industrial scale, which makes them useless in the industrial system. Skills and knowledge are also lost in the process of relying more on industrial agriculture, as practices like seed saving are less widely engaged (this tie between knowledge, culture, and practice will be more discussed in depth in Chapter 7). Also, monocultures prevalent in industrial agriculture can cause problems for local economies, ecosystems, and result in rights abuses for humans and more-than-human beings (Chao 2022).

So, is contemporary farming inherently incompatible with the human agents involved holding caring attitudes towards tomato plants? I think that depends. Though considered impossible to some[6], I contend that an ethic of care, even on the large scale of farming, could possibly be satisfied. It is possible in theory for the landowners, machinery designers, fertilizer producers, pest control developers, and harvesters to have valuing-attitudes, caring relationships, and beneficial consequences for the plants. The scale is what is especially tricky in these cases: at this scale, is all of this possible? This will be interrogated further in the concluding analysis portion of this investigation.

Before moving on to the next context, it is necessary to mention another crucial component of ethics of care in farming: human workers. Wilson argues that the major push towards mechanization in tomato growing coincided with the worker rights movement of the 1960s (Wilson and Gillespie 1999, 205). Within the North American context, migrant farmworkers are a major ethical component of care ethics (Howard and Forin 2019). As moral agents involved with the plants, care ethics might demand affect, certain attitudes, and relational elements like dependency and interactivity. But within the industrial structure, farmworkers are not only on the frontlines of caring for the plants, but also in positions of vulnerability within the larger web of agriculture. Human harvesters work in increasingly dangerous environments, with pesticides and climate change (Klein 2019, 207–33). They are routinely underpaid for their labor and often live in politically precarious conditions. Particularly after the start of the second administration of Donald Trump in the United States, agricultural workers are racially profiled and especially vulnerable to federal threats of disappearing, detention, and deportation (Reichlin-Melnick 2025, ACLU 2025). While this project is primarily focused on human-plant relationships, the context of those particular, cultural, and historical relationships are not confined only to humans and plants. They are conditioned by human interrelationships that may include domination and exploitation, as is present in some farms.

Home Gardening and Tomato Plants[7]

The aesthetics of home gardening might initially be more attractive to care ethicists and environmental philosophers. Home gardening is recognizable by the smaller scale of production and in some cases, the types of tomato plants that gardeners choose as some tomato plants are a better fit for small production than industry. The aim of home gardening is to provide food for an individual, family, or community. Some small farms might qualify as gardens depending on the scale of the operation and the goal. While the goal behind the plant varieties chosen may rely on some similarities with farming (climate, use-able space, etc.) there are other considerations that have a strong influence on tomato plants in home gardens (e.g., foods and uses particular to that home grower, family and community traditions, etc.).

Within my own gardening lineage, the economic market plays a role even within my family's customs of tomato growing. My familial connection with tomato plants comes primarily from my mother's side. My mother comes from a home gardening background and her grandparents on her mother's side ran a dairy farm. The patriarch in that family, Clarence Houston (1905–1999), despite being very frugal, bought his tomato plants as seedlings. One clear reason to buy rather than save seed would be the energy intensiveness of the dairy farm, leaving little time and energy left for the process of saving seed and starting plants from seed.

My mother's father, Joe Driver (1931–2022), was frugal out of necessity. He bought tomato seed and sold seedlings to a local greenhouse, Young's Flowers, in the 1960s and 1970s to add to his family income. Common types of tomato plants he grew were Burpee Big Boys, Beefsteak, and San Marzano. Driver kept some seedlings to plant in his home garden as a family food source. But he didn't save seed. This fits the historical trajectory of tomato plant popularity and culture at the time. Heirlooms were giving way to plants like Burpee Big Boys, which don't work through saving seed because they are hybrids.

In my home, we always bought seedlings. My mother, Val Brelje, thought they were a good investment, because they were not expensive and far less time-consuming than starting from saved seed. Sometimes our tomato plants would re-seed themselves and pop up on their own anyway the following year. Usually, my mother grows Burpee Big Boys, sun golds, yellow pear tomatoes, Roma tomatoes, beefeaters, cherry tomatoes, and more recently, candy tomatoes. For my grandfather, the garden was a necessity; without it, my mother and her siblings likely would not have had enough food to eat. For us, the garden was a habit. We definitely used the bounty the garden produced, but as industrial agriculture has been so widespread throughout my lifetime, I have never considered it a necessity. Even now, when discussing home finances with my husband, as renters, it seems that growing plants is more of an expensive hobby rather than an investment since food in grocery stores can be so cheap, thanks in no small part to industrial agriculture.

There is also market gardening, a phenomenon between home gardening and industrial farming. A great example of this is Rainbow Tomatoes Garden in East Greenville, PA. Dan Wabar, the owner and main farmer, named his company after a garden instead of a farm (Wabar 2021). Why? Partially because of its scale. But the scale is integral to his growing methods. Each season, he knows exactly how many plants are in his acre growing space. Wabar's practice is rooted in his philosophical beliefs. His garden takes one acre of a six-acre property and has the claim to fame of growing 320 tomato varieties. Wabar plants both hybrids and heirlooms, but buys all his seed. This is in part because he wants to ensure that the tomato genetic line is intact when selling at market. He does not produce bulk tomatoes; rather he specializes in producing smaller amounts of many different types of tomatoes.

It seems clear to me that home gardening is, at least at first blush, more attractive to ethical care than industrial agriculture. This is in part due to distance. A closeness between the gardener and the cared-for may not obscure the value of the cared-for in the same way that the distance between owner, tend-er, harvester, consumer and tomato plants in the industrial agricultural context can. Often times, home gardeners put a good deal of time and attention into their plants and might see them as individuals. For some humans, respect and other morally rich attitudes associated with care require knowing and seeing the other as an individual. Witnessing the resource intensiveness of bringing a tomato plant to maturity may instill appreciation for the plant in the gardener.

The relationship between gardener and plant could be less removed, as usually the gardener is the one primarily responsible for the plant.

I worry that this closeness and appreciation could be mistaken for an aesthetic idealism. As I write this, online movements glorifying an idealized domestic femininity and supporting a "return to nature" fill social media feeds. In heavily produced media segments, a (typically young and conventionally attractive) woman appears describing her garden where she preciously and beautifully brings fruit and vegetables into a gorgeous kitchen to process (Kelly 2025). This (arguably political) aestheticization of the care relationship between humans and other humans, along with animals, plants, and ecosystems, is nothing new. And it is an important issue to bring up in connection with care ethics.

Beyond the politicization and aestheticization of home gardening, the tangible process of care in gardening can also go awry. Despite elements in the relationship between a gardener and her tomato plant that are conducive to care, the gardener could fail to hold appropriate valuing-attitudes for their cared-fors. She could fail to provide the right resources for the tomato plant and unintentionally kill it. She could destroy the plant for no reason or uproot it and let the plants die. The dangers of violating the obligations of ethical care are present in home gardening as I will address in the analysis portion of this chapter.

Tomato Plant Cultivation and Seed Saving

Seed saving collectives and small seed companies can be the go-betweens of farmers and home gardeners. Often these organizations aim to amass, catalogue, save, and expand distinct genetic strains of plants through saving seed and experimenting with hybridization. Dr. William Woys Weaver's *Heirloom Vegetable Gardening* (1997) has been credited with the renewed popular interest in heirloom plant varieties. His family's Roughwood Seed Collection is a long-running seed bank. There is a lot of diversity when considering types of seed saving (Nazarea, Rhoades, and Andrews-Swann 2013). Some families save seed of plants that do well in their home garden over generations. As these seeds reliably produce good, consistent plants and fruits year after year, they can be considered heirloom plants. There are also seed banks, where seed is saved as verified strains of plants. Seed companies sell verified strains of seed to consumers on the market. Seed libraries are community organizations that promote gardening by providing seeds free of charge to the community. These seeds are invaluable to those who need them but are not as verified as seeds bought from a seed company or obtained through a seed bank.

While consumption of the fruit is still very important (read LeHoullier's *Epic Tomatoes* and you'll find a buffet of flavors and textures described in wonderful detail), seed saving communities highlight two factors of tomato plant care not front-and-center in the other two contexts: genealogical legacy and lore, and future generations. These are important ways that seed savers can provide care for both humans, seeds, and perhaps the plants that produce them or the future plants that will emerge from them.

Alongside the collection and protection of genetic materials and the stuff of future tomato plant generation, the seed savers often times collect the stories of the families and cultures that have nurtured these plants in the past. Seed savers, by definition, are concerned about the future. The collection of seed is an activity that one does expecting to use the seed again sometime in the future. Sometimes the use is to support a particular gardener, farmer, or producer; other times it preserves a particular family strain of a plant. For tomatoes specifically, there are several resources that detail heirloom varieties. As an example, Carolyn J. Male's *100 Heirloom Tomatoes for the American Garden* (1999) details the stories, origins, and descriptions of a number of tomatoes.

Other times it preserves a particular family strain of a plant that has cultural importance. Truelove Seeds, a small seed company out of Philadelphia, PA, specializes in seeds connected to human diasporas. Their motto is: "Keeping SEEDS is an act of TRUELOVE for our ANCESTORS and our collective FUTURE!" (Truelove Seeds 2025). The ancestors that Truelove is referring to stretch continents and global migrations, as the company emphasizes accessibility to culturally important plants. Within their business model, they include a commitment to seed rematriation (the return of seed to its motherland, or place of origin and cultivation). Seed is more than a mere commodity; it is a connection to the past and future. Truelove Seeds carries a wide variety of tomato plant seeds, from the Russian strain named after Paul Robeson, to the Syrian Homs 11 Tomato and the Al Kuffa Tomato from Iraq (Truelove Seeds 2025). There are also a wide range of varieties from Germany, Greece, and the Americas. This demonstrates the history of the tomato within human communities, traveling across the globe and being tended and passed down in a wide variety of regions and families.

Seed saving is also an activity that enshrines a certain genetic code. An example of this is present in the recent seed rematriation to the Seneca nation by Angela Ferguson, a member of Braiding the Sacred (Braiding the Sacred 2025). In the summer of 2025, Ferguson brought a collection of seeds from the Onondaga nation including a black bean called Burnt House Village Bean, Haudenosaunee Crook Neck Squash, and Seneca Long Ear White Corn (Seneca-Iroquois National Museum 2025). The importance of this rematriation is manifold. Ferguson describes the moral components of her role as a seed keeper:

> Planting is our obligation to Etino'ëh Yöëdzade (our mother, the Earth). When seeds are given, we make a lifetime commitment and intergenerational commitment to keep them going. Planting a seed in the ground is sending it home, like sending a relative on their journey to Sky World. Your obligation as ögwe'ö:weh (Indigenous person) has been completed, let the seed do its job to grow feed the people.
>
> (Seneca-Iroquois National Museum 2025)

Not only is the act of saving seed, growing it, and bringing it back to its original lands an act of care for the seed, it is a moral obligation to earth, as well as to current and future generations of humans. This practice found in the ethics of the Great Law of the Iroquois Confederacy, like care ethics, prioritizes relationships and understands obligations coming forth from those relationships. In some traditions, seeds (and plants) only understand or respond to speaking or songs in a particular language (Coats 2022). As seeds are harvested and processed for preservation (e.g., separated from the fruit, washed, dried and stored), the human caretakers are responsible for speaking and singing to the seeds in these languages to fulfill their obligations of care. Seed saving in this way is keenly tied to not only ethical concerns, but also key practices of culture and identity.

There are also political differences in seed saving organizations and across belief systems. As many gardeners and consumers will know, a single fruit could produce an enormous yield of seeds. In response to this abundance, some understand seed saving as an explicitly anti-capitalist act (Peschard and Randeria 2020, Dow 2021). When you save seed from a fruit and use it to grow plants into the future, you are able to avoid the marketplace practice of buying new seed every year. There may be more variations in saved seed if one is not careful to avoid cross contamination in their fields and gardens.

Others understand it as a scientific cataloguing of a living (or dormant) museum (Conroe 2022). A fascinating larger question that both seed savers and larger horticultural gardeners face: is plant collection compatible with ethical care? If you are cultivating plants merely to collect them (e.g., to have a living museum, of sorts) does that undermine ethical care (Karafyllis 2019)? It might be a type of objectification that undermines the valuing-attitudes that are required in ethical care. This is a wider issue that cannot be covered in this brief section, but it is an important to flag.

Still others consider seed saving compatible with more stereotypical farming: that seed savers can produce hybrids that can serve the wider society as well as individual seed savers (LeHoullier 2021). When considering the application of care ethics in the context of seed savers, valuing-attitudes, relationships that include caring affect, dependency, and responsive interactivity, and successful consequences of caring actions are all at play.

Care in Consumption: How Can We Ethically Care for Something We Consume?

The term "crop" is defined in part by use (Bray et al. 2023). Claiming that any cultivation for the goal of consumption as merely instrumental may not fit in an ethic of care. This initial impulse is understandable, especially given the paradigmatic caring relationship between a human parent and child. It sounds a little too Hansel and Gretel; a gardener lovingly caring for a row of tomato plants while dreaming of slicing the fruit up on sourdough with mayo, like a fairytale witch delighting in fattening up the children she intends to eat. And yet, isn't that practically what the gardener is doing?

As demonstrated above, cultivation and growing of plants for consumption occurs within a wide variety of belief systems and paradigms. Some practices (e.g., native, indigenous, or First Nations) may appear more compatible with care ethics than others (e.g., contemporary industrial scale agriculture) because of the valuing-attitudes espoused in a kinship worldview. Analysis is needed to understand if ethical care is possible across this wide variety of contexts. Returning to the crucial pressure points of care ethics offered in Chapter 3, I want to examine whether the consumption context of tomato plant cultivation can still navigate the following:

1 Attitude that ascribes inherent or intrinsic moral value to the cared-for,
2 Presence of affect, dependency, and interactivity in the relationship between carer and cared-for, and
3 Successful consequences of caring action for the cared-for.

Valuing-Attitudes

I think one of the largest sticking points with farmers, gardeners, and seed savers in living up to the requirements of ethical care is valuing-attitudes. The contemporary ethical context of use and consumption makes this clear: use and consumption must easily fit extrinsic moral valuation over intrinsic. There is usually a firm difference between inherent moral value and extrinsic moral value: the latter is valuable only in so far as it serves the needs or benefit of someone or something with the former. Plants are a commonly assigned extrinsic moral value in anthropocentric communities. If humans are the only things with inherent moral value, then everything else either has no moral value or only extrinsic moral value. This can be used as a justification to promote the growth of some plants, while eradicated others because they are only instrumentally valuable for attaining the ends of humans. For ethical care, however, the carer must have a valuing-attitude that attributes inherent moral value to the cared-for. In most things that humans use merely as a means to an end, these valuing-attitudes are not present, and in some ways, practically cannot be present.

My use of tomato plants as a case study can hypothetically avoid some of the most instrumental treatments of plant life because tomato plants are grown not for the consumption of the plant itself, but for the fruit. So, you can rear tomato plants without your ultimate goal being the destruction or consumption of the plant. Other crops require the plant to be killed in order to harvest the portion that humans eat. For example, carrots (*Daucus carota)* and beets (*Beta vulgaris)* are actually enlarged taproots. Since the taproot is required for the plant to continue living, when the carrot or beet is removed, the plant dies. While the edible parts of broccoli (*Brassica oleracea* var. *Italica)* are the buds at the top of the stem, usually broccoli is harvested by the head, deeply damaging the plant. As buds, the harvesting of Brussels sprouts (*Brassica oleracea)* does not require the killing of the plant, but in industrial settings it is done in a way that does kill the plant. So, even in the case of plants that do not require being killed for their consumption, the means of harvesting may result in their deaths. All of this to say, regardless of whether the crop in question is a fruit or a whole plant, necessarily killed for harvest or not, the very instrumentalization of these plants demands an explanation of whether ethical care could apply or be possible in cases of use. Tomato plants would not be such a presence in the lives of industrial operations, home gardeners, or seed savers were consumption not keenly tied to the plant itself.

Caring Affect

Caring affect regularly plays a role in the relationship: joy when a plant thrives, sorrow when a plant succumbs to some ailment or is not healthy enough to produce fruit. In my interview with Craig LeHoullier, his feeling for his plants is palpable. He speaks about heirloom vegetable plants with passion. And he appears to appreciate the plants for themselves, rather than only for the fruit they produce. As he writes in *Epic Tomatoes*,

> The garden can be a supplier of food, but it can also be a laboratory for experiments and tests, and a museum of living history. All of it is really quite wonderful, and this only begins to express the joy I find in growing tomatoes.
>
> (LeHoullier 2015, 14)

Caring affect was apparent in all the interviews that I did with growers.

Dependency

Tomato growers are deeply aware not only of the dependency of the tomato plants on the growers, but of the grower on the health and yield of the plant. Farmers need the tomato plants to do well for their livelihoods. If the tomato plants do not produce, the human communities around the farmers suffer. Gardeners like Wabar are also depend financially on their tomato plants to do well. Perhaps some contemporary home gardeners do not have the same financial dependence on their tomato plants (though home gardeners who live from their yard might have additional financial costs if the fruit they were planning to survive off of were not available) there are other ways they depend on their plants. For example, when my grandmother was in an assisted living home, my mother provided both plants on her porch (cherry tomatoes) and tomato fruit from my mother's home garden to enrich my grandmother's life and plate. Is it a physical necessity? No. But having fresh, flavorful tomatoes was part of my grandmother's culture and life in a way that she came to depend on.

As stated before, many crops were cultivated by humans over generations to grow in a particular context or geographic region with the support of human intervention. Does this mean the cultivated plant, the crop, *always* depends on constant intervention from the human? Surely not. There are instances, for example, where my family's cherry tomato plant has re-seeded itself from a previous year. Those who enjoy having jack-o-lanterns and compost their seeds or whole pumpkins may be surprised to find thick pumpkin vines growing from the compost heap the following summer. Though, without the initial intervention (i.e., my family planting the cherry tomato seedling in the large pot on the deck, my neighbors buying the pumpkins for decoration that ended up in their compost pile) that re-seeded tomato plant springing from the gravel in the driveway or the pumpkin vines emerging from the compost heap would not be there. Human activity constantly intervenes in the lives of crop plants and their descendants.

Responsive Interactivity

In the case of human-tomato plant relationships, the interactivity of human and plant is straightforward. If a human interacts with the plant, it responds. When the gardener waters the plant, the reaction to the gardener's actions can be seen in a myriad of ways: growth of plant, color of leaves, production of fruit, etc. Changes in the plant elicit responses from the human, as well. The grower may change a fertilization schedule, move the plant to a different area, or harvest fruit depending on the situation. The responses depend on the action and the larger context (environment, genetic make-up, history of that particular plant, etc.), but interactivity between carer and cared-for is present.

Farmers, gardeners, and seed savers all depend on their ability to interpret and provide for the needs of the plant. As Scott Hoffman from Furmano's explains: every year he works to perfect a plan for the season, and every year that plan is adjusted to changing conditions that required preferencing the plants' needs and health. Were these plans not to change, the plants would succumb to illness or climatic stress, and simply die without producing the fruit that they would have been able to if kept healthy. And as passages in *Epic Tomatoes* demonstrate, LeHoullier is exacting in his observation of the effectiveness of his care for his plants. He keeps notes and shares that information with others in the collective.

Beneficial Success Consequences for the Cared-fors

How can tomato plants flourish in a system of use and consumption? I think there are a few options for how this could be fulfilled. One way would be by taking care of the plant such that it grows well. This means successfully fending off pests and providing an environment that supports the growth and health of the plant. This seems clear in current agriculture—carers are deeply committed to the physical success of their plants. Another way that this occurs is possibly through the continuance of the tomato plant's genetic material through time, the continued prioritization of its seeds being given the opportunity to grow.

Conclusion

Care ethics, as described in the first few chapters, is concerned with valuing-attitudes, caring relationships and success conditions being met for the cared-for. Ethics of care espouse the cared-fors as being valuable in and of themselves. Caring affect can help illuminate that value (Little 1995). Its presence in the experience of tomato cultivators can be further evidence of something with moral dimensions occurring in the caring actions provided by humans in the lives of these plants. Connected to them through caring practices, they feel happiness, delight, gratitude, and contentment when the plant flourishes. When the plant responds to the actions of the carer and lives, reproduces, and interacts with its fellow members of the ecosystem (soil microbes, insects, pollinators, fungus, small mammals, etc.), ethical care is apparent. But these characteristics of the caring relationship must happen in tandem with a valuing-attitude.

While the cultivator (gardener, etc.) has the aim of benefiting from their caring actions towards the plant, it needn't mean that the relationship is only extractive. A gardener can have aims that coincide with things that benefit the plant itself as well. Ethical care simply requires certain internal and external commitments: caring affect and valuing attitudes from the carer for the cared-for, and actions provided by the carer that acknowledge the dependency and interactivity of the caring relationship between her and the cared-for.

I do think it is possible for farmers, gardeners, and seed savers to have valuing-attitudes that ascribe inherent moral value to their tomato plants. Whether it is the norm, or something that many or few growers do is not something that I know. But it does seem possible, even at a large scale, for growers to have a valuing-attitude that ascribes inherent moral value to a plant, even if one is also using that plant for one's own ends.

Notes

1 I leave this open to the attitudes ascribing value to the cared-for is because of the great breadth of scientific, religious, and philosophical paradigms that underly particular plant care takers' relationships with plants. There is a huge spectrum of views on inherent moral value as it relates to plants, and my goal in this project is not to defend one source of value against another. Rather, I want to understand if ethical care is possible in the human-plant relationship. For this reason, I commit to a view of non-personhood to allow for less morally robust frameworks of individual practitioners who may still indeed demonstrate ethical care in human-plant relationships.

2 As noted in earlier chapters, she favors relational ethics to care ethics because care ethics is too bound up with emotion for her and may skew moral obligations because of particular emotional ties or responses to individuals. So, while direct adoption of her contextual ethics into an ethic of care might be frowned upon, as she establishes her own view as an alternative to care ethics, her approach is not completely at odds with an ethic of care. At least, in my case because unlike Kittay, I am focusing on relational connections rather than virtue ethics as a normative framework.

3 A previous version of this chapter called this practice "Industrial Agriculture" instead of "Farming." I think "Farming" is a more appropriate title. The structure of industrial agriculture still utilizes small to medium farms as a source of produce for their distribution or processing. As Dr Jayson Harper explains, contemporary farmers (and farms) are usually contractors to a larger industrial firm. And "industrial agriculture" can sometimes be an ethical non-starter for some folks. It does have a divisive ring to it in contemporary discussions about plant rearing, so, after my discussion with Dr Harper, I decided that "farming" would be a better fit for my purposes and the actual case study I'm trying to investigate than "industrial agriculture." It is important to note that some of the people I interviewed who were quite critical of industrial agriculture. They would probably also be critical of it by another name. But perhaps other readers would not have an instantaneous negative association with farming like they can have with industrial agriculture. Additionally, Dan Wabar, who shows up in the gardening section, is a producer of tomatoes for the market, but intentionally calls his operation a garden rather than farm. He says this is in part because of the difference in scale—a major component of this section as I was working with the terminology of industrial agriculture—and because he considered the scale crucial to the care he was able to give his plants. His use of the differentiation of farm and garden seems to hold as a reasonable differentiation of context—without needing to add "industrial."

4 Dr Weaver prefers the term "heritage" over "heirloom" because in translation into other languages than English, the term "heritage" is a better term. But "heirloom" has caught on in the U.S. context, and especially when discussing tomato plants.

5 It should be noted that practitioners interviewed with this paper do not agree unanimously that plants are not sentient. Dan Wabar, discussed in the following section on gardening, notes that all life requires suffering. And at some point, in an effort to eliminate suffering, he sought to become a fruitarian—because he believes that killing a plant for consumption does create suffering in the plant. However, as he found this biologically unsustainable for human life to live only off of fruit and seeds, he shifted to try to reduce overall suffering—admitting that in the consumption of plants for food, there is still suffering on the part of the plant.

6 Dr William Woys Weaver, who plays more of a prominent role in the section on seed saving, is an animist. He believes that all living things have a spirit or soul. While suffering of spirit-filled plants did not come up, he did state his belief that industrial farming is "soulless." Both Wabar and Weaver also criticize industrial farming as extractive of the soil (or land more generally).

7 In the context of urban spaces, Marcello Di Paola (2017, 37) distinguishes gardens from "peri-urban agricultural lands" in part by allowing for gardens to be for more uses than mere production. However, since this chapter is focused on tomato plants connected to consumption, I am more interested in differentiating industrial agriculture and farms from gardens by size. While hypothetically people could grow tomato plants for reasons other than to consume the fruits, the vast majority of current cultivation revolves around the consumption. So that is my focus in this chapter.

References

American Civil Liberties Union Southern California. "U.S. Supreme Court Grants Stay in L.A. Raids Case." www.aclusocal.org/en/press-releases/us-supreme-court-grants-stay-la-raids-case

Bartram, John & Son (Philadelphia) and Henry G. Gilbert Nursery and Seed Trade Catalog Collection. *A Catalogue of Trees, Shrubs and Herbaceous Plants, Indigenous to the United States of America*. Bartram and Reynolds, 1807. http://archive.org/details/catalogueoftrees1807john

Blakemore, Erin. "The Neolithic Revolution—Facts and Information." *National Geographic*, April 4, 2019. www.nationalgeographic.com/culture/article/neolithic-agricultural-revolution
"Braiding the Sacred: A Movement of Indigenous Corn Cultures." https://braidingthesacred.org/home.
Bray, Francesca, Barbara Hahn, John Bosco Lourdusamy, and Taigo Saraiva. *Moving Crops and the Scales of History.* Yale University Press, 2023.
Chao, Sophie. *In the Shadow of the Palms: More-Than-Human-Beings in West Papua.* Duke University Press, 2022.
Coats, Waylon. Interview. 2022.
Cockshaw, Rory. "The End of Factory Farming: Alternatives to Improve Sustainability, Safety, and Health." *Voices in Bioethics*, 7 (2021).
Conroe, Allison. "How Saving Seeds Could Save The Planet." Learn—GlobalGiving (blog). www.globalgiving.org/learn/indigenous-seed-saving/ (accessed May 31, 2022).
Curry, Helen Anne. *Endangered Maize: Industrial Agriculture and the Crisis of Extinction.* University of California Press, 2022.
DeWitt, David. *Precious Cargo: How Foods from the Americas Changed the World.* Counterpoint Press, 2015.
Di Paola, Marcello. *Ethics and Politics of the Built Environment.* Springer, 2018.
Dow, Katharine. "Bloody Marvels: In Situ Seed Saving and Intergenerational Malleability." *Medical Anthropology Quarterly*, 35(4) (2021), 493–510.
Eggers, Robert. *The Witch.* A24. 2016.
Harper, Jayson. Interview. 2021.
Hoffman, Scott. Interview. 2021.
Howard, Neil and Roberto Forin "Migrant workers, 'modern slavery' and the politics of representation in Italian tomato production." *Economy and Society*, 48(4) (2019), 579–601.
Karafyllis, Nicole Christinc (ed.). *Theorien der Lebendsammlung* [Theories of Living Collection]. Varlag Karl Alber, 2019.
Keller, Evelyn Fox. *A Feeling for the Organism: The Life and Work of Barbara McClintock*, 10th anniversary edn. Henry Hold, 2003.
Kelly, Annie. "Truly, Tradly, Deeply." Podcast series. Cursed Media, 2025.
Klein, Naomi. "Season of Smoke." In *On Fire: The (Burning) Case for a Green New Deal* (pp. 207–233). Simon & Schuster, 2019.
LeHoullier, Craig. *Epic Tomatoes: How to Select & Grow the Best Varieties of All Time.* Storey Publishing, 2015.
Li, Alice M. "Ecological Determinants of Health: Food and Environment on Human Health." *Environmental Science and Pollution Research*, 24(10) (2017), 9002–9015.
Little, Margaret Olivia. "Seeing and Caring: The Role of Affect in Feminist Moral Epistemology." *Hypatia*, 10(3) (1995), 117–137.
Longino, Helen E. *Science as Social Knowledge: Values and Objectivity in Scientific Inquiry.* Princeton University Press, 1990.
Lundqvist, Jan, "Avert Looming Hydrocide." *Ambio*, 27(6) (1998), 428–433.
Monticello. "Tomato - Solanum Lycopersicum Cv." www.monticello.org/house-gardens/in-bloom-at-monticello/tomato/ (accessed August 23, 2022).
Nazarea, Virginia D., Robert E. Rhoades, and Jenna Andrews-Swann (eds). *Seeds of Resistance, Seeds of Hope: Place and Agency in the Conservation of Biodiversity.* University of Arizona Press, 2013.

Norwood, Vera. *Made from this Earth: American Women and Nature.* University of North Carolina Press, 1993.

Palmer, Clare. *Animal Ethics in Context.* Columbia University Press, 2010.

Peck, Steven L. "Death and the Agricultural Process." *Agriculture and Human Values*, 27(1) (2010): 105–109.

Peschard, Karine and Shalini Randeria, "'Keeping Seeds in Our Hands': The Rise of Seed Activism." *The Journal of Peasant Studies*, 47(4) (June 6, 2020): 613–647.

Reichlin-Melnick, Aaron. "Trump's Immigration Actions Are Taking a Toll on Local Economies – Here's What the Data Says So Far." American Immigration Council Blog. www.americanimmigrationcouncil.org/blog/immigration-toll-on-local-economies-what-the-data-says/ (accessed August 27, 2025).

Rollin, Bernard. "Confinement Agriculture from a Moral Perspective." In *The Routledge Handbook of Food Ethics* (pp. 253–263). Routledge, 2016.

Seneca-Iroquois National Museum. "Seneca Seed Rematriation." https://youtu.be/J0oobj5OSc8?si=rcCk_v2qyv6Ddy7w.

Smith, Andrew F. *The Tomato in America: Early History, Culture, and Cookery.* University of Illinois Press, 2001.

Truelove Seeds. https://trueloveseeds.com/collections/vegetables-and-grains/a-z-type_tomato.

Vodouhè, R. and A. Dansi. "The 'Bringing into Cultivation' Phase of Plant Domestication Process and Its Contributions to In Situ Conservation of Genetic Resources in Benin." *The Scientific World Journal* (2012).

Wabar, Dan. Interview. 2021.

Weaver, William Woys. Interview. 2021.

Wilson, David Scofield, and Angus K. Gillespie. *Rooted in America: Foodlore of Popular Fruits and Vegetables.* University of Tennessee Press, 1999.

7 Guarding Giants

The Ethics of Caring for Giant Sequoia Trees in Wild Contexts

Caring for Wild Plants

Ethics of care has an air of domesticity to it. Grounding initial theories in the relationship between parents and children intimates this sense that care is something that occurs within the spheres of human families, with the home as setting. What role, then, can care ethics have in areas outside of human domesticity in places commonly defined, in part, by the absence of human presence or interference? There are many philosophical definitions of "wildness" and "wilderness" within environmental ethics. This debate over what separates the wild from the non-wild looms large. Defending a particular robust definition of wild-ness is not my goal here. Rather, I am interested in a case study that examines caring relationships between humans and wild plants. To do that, while largely bracketing the ontological "what is wild?" question, I hope my choice of a popular, naturally occurring and governmentally protected tree will suffice. I am less concerned with the strict definition of what qualifies as a wild plant, and more interested in how human relationships form (or fail to form) with these plants. Human presence or absence is a difficult part of care ethics with wild plants. In this chapter, I apply care ethics to the case study of giant sequoia trees.

As this case study will reveal, there are wildly different world views that have an impact on what caring actions are permissible or even obligatory. Beyond that application, I examine the role of humans more broadly in the lives of wild plants. There are alternative ways to understand the role of humans in the natural world. By centering relationships, care ethics requires us to closely analyze the types of relationships we have with natural entities. How we understand and describe these relationships is useful for understanding whether the components of care ethics even apply in the context of wild plants. For example, contrast the tomato plants in a home garden in the previous chapter and a grove of trees miles away from any homestead or home garden. The home garden is regularly tended, in part because the human carers placed those plants in that particular spot and are now in some way needed to maintain the life and growth of the plant. The carers intervene in times of drought by implementing irrigation or adding water to the wilting

DOI: 10.4324/9781003603665-8

plants. The plants can be highly dependent on the actions of the carers, especially in the case of plants like tomatoes, which have been selectively bred and removed from native contexts overtime. Wild groves of trees located far from human settlements, by contrast, may be more or less left alone. They are there due to a collection of other forces and factors rather than human choice or agency. So, their survival traditionally does not rely on much intervention from human agents.

Within such a context there's reasonable uncertainty about the appropriateness or even possibility of care. Of particular concern is the importance of dependence and responsive interactivity. Dependence could be understood in a number of ways. Within the context of human-plant relationships, one might understand it as the need for intervention. The world outside of human influence simply exists in its cycles, moving and managing without any need for humans or their actions. Some, like Eric Katz (2023), consider instances of interaction between humans and this natural (i.e., non-human) world to be bad or wrong which might qualify as a bad intervention on an otherwise autonomous system. Others, however, consider these interventions to be useful and necessary actions that improve or help the system. For example, let's say that humans have removed apex predators from an area where they historically managed the population of deer. Without the apex predators, deer populations have exploded and are overgrazing forest land, putting the ecosystem off kilter. Human intervention by hunting deer during a scientifically determined season may help put the ecosystem back into balance. Some might even consider this to be a type of appropriate restoration work that the ecosystem deserves because of the previous human intervention (e.g., removal of wolves) resulted in this imbalance. This results in the ecosystem being dependent on the continued intervention of humans to manage the deer population. As it relates to plants, the overgrazing by deer may threaten wild plant populations. Had human actions not resulted in the removal of apex predators, the continued intervention of killing deer to manage overpopulation would not be required. So, wild plants may be dependent on human intervention in restoring or managing imbalances produced by previous or historical human actions that contributed to the imbalance. Perhaps even in the case of non-human caused imbalance (e.g., a natural disaster unrelated to anthropogenic climate change), humans would be required to intervene for the sake of upholding the Land Ethic (Leopold 1949). This kind of dependence could also be understood as interdependency that includes tending without domestication. This is roughly what the National Park and Forest services could be said to do.

Dependence could also refer to something's need for survival, like how humans are dependent on plants for their continued existence. This notion is vaguely invoked when referring to the Amazon rainforest as the lungs of the Earth. The implication is that everything (humans included) relies on the activity of the plants within the Amazon rainforest. Were the Amazon rainforest to be dramatically changed, it would be like a person's lungs failing to function well. It could even lead the to the death of the human, or by comparison, the death of living beings on Earth.

There is a different paradigm that emerges from native, indigenous, and First Nations traditions. Dependence could refer to an interdependency that is not only physical but also social, religious, or cultural. Humans as social creatures rely greatly on our social relations for strictly survival-oriented needs (e.g., caretakers teaching small children not to touch the hot stove) and for emotional, psychological, and social ones (e.g., learning healthy expression of emotions, developing the right amount of self-esteem, and learning how to play well with others and benefit from friendship). The perceived lack of language on the part of plants in relationships with humans may suggest that plants are unable to fulfill any social needs a human might have. But for many individuals and traditions, plant communication is real and impactful in their lives (Molina and Shorter 2025). Additionally, it is the cultural or moral obligation of humans to provide gifts or services to plants. For example, in many indigenous traditions, humans are required to practice ceremonies (including songs and dances) to maintain balance in the more-than-human world (Molina and Shorter 2025). This is the case even with wild plants, with whom (or which) humans have a relationship, even if it is not an intensive relationship (e.g., placing them in a certain environment to tend, as described in the previous chapter). In this way, both plants and humans are dependent on one another for the gifts that they give in relationship to maintain a balance in the natural world (Kimmerer 2013).

Contemporary practices, like social media production and consumption, make dependency claims more complicated, particularly when we consider parasocial relationships with plants. Some relationships that we feel we have with plants might be very one sided. Imagine you follow a social media account where a human documents a giant sequoia tree daily. Every day, you look at the picture and read the update. You start to develop a relationship with that tree parasocially—i.e., you interact with it online through media—but the tree and even the person posting about the tree are unaware of you. This could just as easily happen with bonsai trees and tomato plants posted online. But the charisma of some plants, like giant sequoia trees, make the prospect of being in a relationship with them through social media all the more tempting. This is a threat to the responsive interactivity (or bolder, reciprocity) that care ethics requires. If the relationship is so one-sided that only one party is aware of the relationship, it may not qualify as a caring relationship. The parasocial relationship contrasts the relationships developed in native, indigenous, and First Nations, where plants are active agents in direct contact with humans though physical, spiritual, and cultural means.

These different senses of dependency will be clear in the results from the interviews analyzed later in this chapter. But for now, it is worth flagging the possibilities of dependency as we embark on an examination of human-sequoia relationships. While some of the interviewees note a lack of this type of interdependence between giant sequoias and human beings, even they admit exceptions—particularly in the case of climate change. The activities of humans are affecting wild plants, like the giant sequoias, to the point that in some ways the plants are dependent on humans discontinuing behaviors that continue to contribute to a warming, changing, and chaotic climate. But this dependence is bracketed as only due to the radically bad actions of humans. Under less extreme conditions, some paradigms support the view that wild plants like the giant sequoias, would have little to no need for us or our interventions.

Giant sequoias are iconic trees with a rich history. In what follows, I argue that human moral agents can be in ethically caring relationships with giant sequoia trees. I hold that in the human-sequoia relationship, humans can have valuing-attitudes towards giant sequoia trees, have relationships with them that include caring affect, dependency, and responsive interactivity, and that the actions of human carers can benefit or hinder the well-being of the giant sequoia cared-fors.

These trees have a reputation for inspiring awe in the humans who visit and interact with them (Tweed 2016, xv–xvii). They are associated with American identity in some ways. For example, at present, they only naturally occur within the U.S. Giant sequoias have physical dimensions and life spans that puzzle and excite viewers. I first interacted with giant sequoias as a girl in Yosemite National Park. I remember the smell of the forest, and my inability to really understand what I was seeing. Perhaps in Kantian terms, my imagination was foiled in its attempt to unify the image of the unfathomably large tree, as my reason struggled to "tame" it. Perhaps it was an experience of sublimity. The trees are challenging to capture adequately in a photograph.

I chose giant sequoias for this case study because they are one of the wild plants that stick out in my memories. While I could have chosen other wild plants associated with fond memories (e.g., my mom pointing out ditch weed on a trip to the country to collect plants on the side of the road or the pine tree groves that my father's mother loved), giant sequoia trees have a greater general symbolism within the U.S. I did not want to choose a wild plant that few know about or that is primarily valued because of human consumption or use. And, in keeping with the themes of the previous two case studies, I wanted it to be a plant that has some connection to my life.

Throughout this chapter, I support my thesis that human-sequoia relationships can count as ethically caring relationships because they account for attitudes of intrinsic moral value, relationships with caring affect, dependence, and responsive interactivity, and beneficial outcomes for the cared-fors. First, I address the historical and contemporary context of these relationships and then provide insights from contemporary caretakers of giant sequoia trees. These contemporary caretakers break down into two primary groups: National and State Parks Service employees and Native local peoples. Because of the historical "setting-apart" of these trees, and their now protected status, the range of people who have close, constant, and enduring caring relationships with these trees is limited. However, as I will address near the close of this case study, climate change puts pressure on the role that those outside this intimate proximity play in caring for these plants. And indeed, many people who treasure these plants hope that the preservation of sequoia trees can be a rallying point for more people to push for policies that urgently address climate change.

Historical Overview and Contemporary Contexts

Today, they grow in a limited space on earth's surface, occurring in California between Placer County, located north of Sacramento and Yosemite National Park, and Sequoia National Forest, located northwest of Bakersfield (Save the Redwoods League). They grow in about 75 groves (Save the Redwoods League), generally between 4,590 and 7,050ft elevation (Fire Effects Information System). The lands on which giant sequoias grow are mostly owned by the United States federal and California state governments. Thirty-three groves are on land managed by the Forestry Service under the United States Department of Agriculture (USDA 2022). Three groves are located in Yosemite National Park (Dickman 2022). In Sequoia and Kings Canyon National Park, there are about 40 groves (U.S. National Parks Service 2022a). Three groves are on Tule River Tribe land (Willard n.d.) and one grove is in Alder Creek, private land that the Save the Redwoods League has arranged to buy (Benson 2022). Given the limited number of giant sequoias, access to them can be greatly restricted.

Geological evidence, by the way of fossils, suggests that "ancestral forms" of *Sequoiadendron gigantea* extended far beyond its current territory (Hartesveldt 2005, 15). Climatic changes in moisture and temperature likely reduced their range (15). Today, giant sequoias live in an environment with high moisture and deep soils, with some living on plateaus, some in cliff areas, some in valleys (Demetry 2022).

Native histories with giant sequoias vary by tribe. In some cases, histories may be difficult to access due to a legacy of colonialism, which both leaves native histories and native contemporary experiences out of broader narratives about cultural items in the United States. As the National Park System acknowledges, "The United States Federal Government, along with local settler governments, have histories of forced movement from areas where giant sequoias grow. For example, when Yosemite National Park was established, native Miwok people were forcibly removed to "preserve" the land" (Yosemite National Park). Of course, Yosemite had been tended to over generations by Southern Miwoks (Spence 1996).

What makes the fiction that Yosemite is pristine (meaning untouched by human influence) that much more painful is the restriction of native peoples from practicing their ways with giant sequoias and other plants, animals, and ecosystems. Native, indigenous, and First Nations peoples[1] who have a long history with the trees have a tumultuous history regarding access to them, special to many area tribes (some federally and state recognized, others not). Some of the stories have been lost over generations of forced relocation and cultural genocide. Some cultural practices have been disallowed because of the protected status of the sequoias. This restriction continues today, though there is more cooperation between federal and local governments and native leadership to co-steward lands that include giant sequoias (University of Washington School of Law 2024). This is in part due to the catastrophic fires that have ravaged the limited land on which giant sequoias live. Traditional ecological knowledge (TEK) is a catch all term for knowledge and practices emerging from native, indigenous, and First Nations communities related to the environment or ecologies. Many native, indigenous, and First Nations communities have particular practices that have been developed over generations to live with and in the environments where they are located. One prominent collection of TEK revolves around fire. As Elspeth Hays notes in her research into trees that produce food and their historic uses, she finds that spaces she once considered wild (i.e., untouched by humans) were actually tended by native, indigenous, and First Nations peoples, often through the use of fire (Anderson 2013). Ethnographer Kat Anderson extensively documents management practices and TEK in the context of modern-day California (Anderson 2013). Fire is one of the primary methods of land management alongside irrigation, transplanting, pruning, and others (Anderson 2013, 351).

Despite this, it is important to mention the difficulty of establishing what those early relationships were like. While some practices remain, e.g., speaking prayers in sequoia groves in native language to be delivered into the heavens (Coats 2022), there are also instances of difficulty finding words for the giant sequoia trees in languages that belong to native, indigenous, and First Nations groups whose language has been systematically destroyed by settler, colonial, and dominant non-native, non-indigenous, and non-First Nations groups (Interviewee A). Language is bound up with cultural knowledge and practice (Jacobs 2025). Sometimes, if the practice ceases, the language and cultural knowledge connected to the components is lost. So, if access to a particular plant, location, or ecosystem is restricted, then practices that involve that plant, space, or system can slowly be lost. And so it is unclear what ceremonies, words, songs, or activities with the giant sequoia trees would have been present in pre-contact times.

If these words and practices do still exist in small corners of groups despite the long history of cultural and linguistic eradication campaigns, it would not be surprising that accessing that knowledge would be difficult for outsiders. This is a very important point, not only in the context of care ethics, but ethics and normative case studies more broadly. I hold the view that not everything is for everybody. By this I mean that, especially in the context of academia and research through the university system (and other public and private institutions), relationships between researchers and native, indigenous, and First Nations people can be highly extractive. As Prue Gibson details in *The Plant Thieves* (2023), there are instances where institutions restrict releasing information to different audiences because the information can be misused and abused in such a way that threatens or eradicates the things being identified, documented, and described. For example, an institution might know the location of psychotropic plants in the wild and release that information only to indigenous groups who use the plants in cultural and religious ceremonies. Were anyone to be allowed to access that information, poachers could find the plants and over collect them in order to profit. In cases like these, it makes sense that not everyone has access to all of the information.

Naming plants is also fraught and reflective of the relationships we have with them (Gibson 2023). The desire to know is strong. The names of plants can be rich and illuminate an entire history, context, and use that would otherwise not be known if not for the names of these plants. Names can reflect identities, uses, contexts, values, and connections. If the name of a plant in an indigenous language holds that cultural knowledge (e.g., how the plant is used, or what position within the community it has), I completely understand why there would be hesitancy to share that information with an outsider to be recorded and held apart from the use and context of the community from and in which it originated. Sometimes wisdom (especially relevant within the context of ethics) requires a tempering of knowledge extraction in consideration to the context the knowledge is produced and what it will be used for outside of that context. There may not be a good enough reason to extract the information from the community.

Within the U.S., settler history contributes to the foundation of contemporary practices. It is contested which settler first "discovered" giant sequoias (Hartesveldt 2005, 2–4). Given the previous conversation on Native history, it should be clear that none of these interactions constitute an actual human discovery of a tree, because Native and indigenous groups had interacted with the trees for generations. Rather, settler interactions signaled a start to our contemporary non-Native relationship with giant sequoias. Within the colonial myths of discovery, some settlers were resistant to believe that the giant trees were real (Tweed 2016, 4). Eventually, they believed but needed to share their "discoveries" with others. Unfortunately, in the pursuit of spreading the word about these hard to believe trees, people required their destruction. For example, some wanted to cut down the trees to put them on display in other parts of the country and world (5). The bark was stripped off from one tree, named The Mother of the Forest, to put on display in the Crystal Palace in New York City (9–11). How appropriate that the Mother was stripped of her bark, divided into chunks, and shipped for an oddities display. Others used the trees as a source of recreation. In 1856, the Mammoth Grove Hotel was established in the Calaveras Grove (modern day Calaveras Big Tree State Park) as a camping area for travelers and sight seers (14–15).

Interestingly, the strict governmental protection of the giant sequoias was prompted in part by this settler abuse of giant sequoia trees. For example, between the mid-19th and early 20th century giant sequoias were harvested for display and the production of goods. However, the material of giant sequoias is not that useful for industrial use because the wood was often brittle and would splinter (Hartesveldt 2005, 9).

Not all settlers were on board with using the trees by killing them or making them the site of recreation. The philosophical, political, and spiritual impulses of preservationists informed policy and law formation that protected these trees (Tweed 2016, 109–10). Perhaps most famous in the Western cultural memory is John Muir, who fell in love with the trees (61). Muir writes,

> The Big Tree (*Sequoia gigantea*) is Nature's forest masterpiece, and, so far as I know, the greatest of living things. It belongs to an ancient stock, as its remains in old rocks show, and has a strange air of other days about it, a thoroughbred look inherited from the long ago-the auld lang syne of trees.
> (Muir 1901)

This love inspired the creation of a National Park. The second National Park in the United States, Sequoia National Park, established in 1890, is the direct result of humans valuing giant sequoia trees in and of themselves. According to the Sequoia and Kings Canyon National Park (SEKI), it "was the first national park formed to protect a living organism," its namesake (U.S. National Parks Service 2022b). The only other prior existing National Park, Yellowstone National Park, was established in 1872. Yosemite National Park was established the same year as Sequoia National Park.

The establishment of these parks entailed an expulsion of native and indigenous caretakers over time, evident in this Petition to Congress from the Yosemite, Mono, and Piute Indians circa 1890:

> We willingly keep the peace, we have no desire to do otherwise, but it is with an uneasiness that we see the time approaching when we must leave this spot which has been the home of our people from time immemorial.
> (Castillo 1978, 275)

Despite living in the area for around 4,000 years (and the area being inhabited for possibly 8,000 years), native and indigenous communities were removed and the management taken over by government systems (Yosemite National Park).

It is important to note that the struggle for proper management, treatment, or relationship with the lands (including the groves of giant sequoia trees) is ongoing. There is evidence that even in the aftermath of this takeover of the Yosemite valley, native and indigenous groups petitioned the government with concerns that could have been written today.

> Soon after this valley was taken away from us by the whites, the great Washington Council gave it to all the white American people for a pleasure ground, a park, where they might come and see the great rocks and waterfalls, and enjoy themselves. Now it seems to us that the laws

> imposed upon the head men of this valley by the Washington Government are being wilfully disregarded, and that Yosemite is no longer a State or National Park, but merely a hay-farm and cattle range.
>
> (Castillo 1978, 275)

Currently, native, indigenous, and First Nations communities are struggling worldwide to regain land that has been exploited for profit (NDN Collective). This includes areas of California where some state funding has been used to restore ownership of lands to native and indigenous groups (California Grants Portal). Despite successes like these, the United States Federal Government is again threatening to privatize public parks lands, allowing those with financial means to purchase the land and use any resources it determines useful on it (McConville, Lutz, and Rowland-Shea 2025).

Government owned lands vary in their goals and uses. Federally owned lands, state owned lands, and locally (county or city governments) owned lands have different protocols, which can have an impact on entry fees and permits, and even allowed uses. On the national level, public lands (i.e., lands owned by the government) generally tend towards one of these two ends: preservation or conservation. Preservation is an approach towards land management that aims to limit human interference with ecosystems. In practice, this might look like the government restricting access to the land, prohibiting non-native animals or plants on parks lands (that includes dogs), and disallowing industries (like timber, oil, gas, water, and mineral) to extract parts of the ecosystem deemed resources. There are a range of philosophical positions that support preservation; for example, as mentioned in Chapter 2, Paul Taylor's work on a respect for all living beings with a telos, or natural goal towards which the organism can pursue and in which it can flourish (Taylor 2011). Conservation as an approach aims for humans to live within the land. Aldo Leopold's Land Ethic is probably one of the most famous examples of this view, as Leopold argues that humans are a part of the ecosystem and must aim to live in a way that supports the aesthetic quality and biological stability of the "biotic community" (Leopold 1949). National forests more frequently engage in a conservation approach to natural resource use (Westover 2016). So, logging is not rare in National Forests, while it would be highly unusual in National Parks. There are giant sequoia groves in Sequoia National Forest. Controversial logging plans have been put forward for the forest, which may endanger the 33 Giant Sequoia groves in the Forest (Kohlruss 2024).

Giant sequoias now live in several groves throughout current-day. Since the landowner decides what happens to the trees, the Save the Redwoods League works to transfer privately owned land with giant sequoia trees into nonprofit and public ownership to ensure that the trees are protected. Two primary contemporary relational care contexts with humans and giant sequoia trees are governmental and local non-profit services and native community members. Part of the reason why the care communities are much smaller for giant sequoias is because of their relatively small numbers compared to bonsai trees and tomato plants.

Within the National Parks and National Forest contexts, I spoke with several employees including William C. Tweed, a retired NPS employee and prolific author on Giant Sequoias; Garrett Dickman, a botanist at Yosemite National Park; Athena Demetry, currently at Yosemite National Park and formerly instrumental in sequoia habitat restoration projects at SEKI; Teresa Benson at the Sequoia National Forest; and Anthony Ambrose of The Marmot Society, a nonprofit that works with the NPS. I also spoke with three local Native people about their relationships with giant sequoias, including Interviewee A, Kerri Vera, Director Department of Environmental Protection, Tule River Tribe, and Waylon Coats, vice chairman of the Southern Sierra Miwuk tribe.

Application and Analysis

In my discussions with contemporary practitioners, we addressed the three major components of a plant-inclusive care ethic: attitudes that ascribe intrinsic or inherent moral value to the cared-for, caring relationships that include caring affect, dependence, and responsive interactivity, and consequences of caring actions that can benefit the cared-for. Similar to the prior two case studies, this context (that of wild trees) presents a unique set of issues within this framework. With bonsai, the main concern was whether the bonsai process could even benefit the plant. For tomato plants, the primary concern was whether caretakers could really have attitudes that ascribe intrinsic or inherent moral value to the plants that produce crops. For this wild tree context, as you'll see in my analysis below, proper attitudes and beneficial consequences are clearly demonstrable. But when it comes to the caring relationship between humans and giant sequoia trees, caretakers did not agree on dependence and responsive interactivity.

Valuing-Attitudes

Beginning with attitudes that ascribe intrinsic or inherent moral value, carers were quick to acknowledge that giant sequoia trees have value. Awe, as an attitude, was a widely held attitude towards giant sequoia trees by interviewees. Awe could be interpreted in a strictly aesthetic way. But so often, colloquially and technically, awe can connect with divinity and the sublime (for Kant, an aesthetic judgment, but one that engages reason and the capacity of the human mind to assert itself against the unfathomable). Both divinity and sublimity, to me, suggest something involving inherent moral value. If I am in awe of something, that something is importantly different, set aside as something that can inspire such a strong response from me. Muir's descriptions of natural lands and the giant sequoias evoke divinity. Importantly, awe does not stand alone as a moral foundation for the caring relationship; it is merely one attitude that suggests that carers value the giant sequoias inherently.

Some interviewees moved beyond awe to love, respect, and kinship. One described being taught to love nature by their grandmother (Vera 2022). As mentioned before, love suggests having inherent moral value for the entity that is loved. Two interviewees also identified attitudes of respect, one mentioning the right of the wild tress to be here (Coats 2022, Vera 2022). One also held an attitude of kinship for the giant sequoias, explaining that giant sequoias were considered grandparent figures within their community (Coats 2022). This clearly demonstrates a view of inherent moral value for giant sequoias.

Unlike interviews in either the tomato plant or bonsai chapters, some interviewees who I talked with were quick to distinguish different ways in which giant sequoias were valuable. Many first discussed the value of the trees in relation to their ecosystem and broader types of ecosystems throughout the United States. The emphasis on trees as organisms within an ecosystem stood out as a key way in which a giant sequoia could be valuable. While giant sequoia bark can be useful in traditional ecological knowledge (TEK), uselessness of giant sequoia for settler industry and timber was a theme (Coats 2022). One interviewee even flagged that one possible reason for the mystery surrounding giant sequoias might be their uselessness for industry. Giant sequoia timber (Hartesveldt 2005, 55) tends to shatter when harvested, so much of the tree is not useful for making planks or other solid wood slabs when the tree is felled (Dickman 2022). Because much scientific inquiry can be driven by profits and motivated by utility, an industrially useless tree likely won't be the first focus of research dollars. Additionally, the interviewee continued, the age and rarity of the trees makes scientists slow to embark on invasive investigations (Dickman 2022). Parks interviewees also noted the relationally impressive size and age of these trees. Perhaps their dimensions make them more interesting to humans. Regardless, the interviewees' attitudes suggest that they consider giant sequoias to be inherently morally valuable.

Caring Affect

Affect, or feeling, is a common aspect of the human-giant sequoia relationships on the part of the carer. Awe appears again as a commonly described attitude and emotion of the people I spoke to about their work with giant sequoias. Perhaps surprisingly, even one interviewee noted that even though they have worked with giant sequoias most every day for some time, they still experience awe when entering a grove (Tweed 2021). They also report park visitors responding with awe when they encounter the trees. Another interviewee reports feeling mesmerized by giant sequoia trees and feeling the life in them (Coats 2022).

I think something more powerful forms of caring affect in the relationships between carers and trees are evident when examining feelings related to the death of giant sequoia trees. "I've counted a lot of dead giant sequoias, and I don't like it. It's like counting dead people. It clobbered me" (Dickman 2022). This quote from an interviewee appeared in the New York Times soon after our interview and reflects the depth of feeling that care takers have towards the trees. The death of giant sequoias prompts the feeling of heartbreak in more than one interviewee (Vera 2022, Benson 2022). There is a deep sense of loss when they witness the death of these large trees.

Dependency

Dependency between carer and cared-for in the case of giant sequoias is less clear. Some interviewees were reluctant to admit any dependency between humans and giant sequoias (Dickman 2022). One admitted that they now have a reliance on giant sequoia trees because the trees have had such an impact on their life and are now integrated into their community. But had this person not interacted with the trees in the first place, they would not feel any dependency on the trees (Demetry 2022). I take this to be a type of social dependence rather than physical dependence. Two respondents answered from what appeared to be more of a scientific understanding of dependency; in that, humans are not directly dependent on the ecosystems in which giant sequoias live. Another interviewee describes the shift in the dependence relationship between humans and giant sequoias historically (Vera 2022). At a time when things were more in balance, this interviewee said that giant sequoias did not depend on humans. But with the rise of catastrophic fires, there is a need for humans to intervene on behalf of the trees (Vera 2022).

This also seems compatible with what one interviewee flagged as the pragmatic value of re-envisioning the dependency within relationships between humans and giant sequoias. According to this interviewee, humans should think as though they depend on sequoia trees, because sequoias are important gauges of environmental health (Benson 2022). Another interviewee describes this phenomenon in terms of giant sequoia trees being an indicator species (Vera 2022). Sequoia trees respond to environmental changes and pressures, especially useful for us to be aware of in this age of dramatic climate change.

Another interviewee mentioned a mutually beneficial but complex relationship between humans and giant sequoia trees (Coats 2022). In this interviewee's context, the community practices prescribed burns to support the giant sequoia trees and their ecosystems. The community also depends on the giant sequoia trees for their cultural identity through creation stories (Coats 2022). In this way, the trees and humans are in a mutually dependent relationship.

Putting aside strict biological definitions of dependence, a re-envisioned sequoia dependency on humans, if anything, as one interviewee noted, giant sequoias are dependent on us "making good decisions" (Benson 2022). They need us to stop contributing to climate change through greenhouse gas emissions so that the trees' ecosystems continue to support them. And more positively, they may depend on us to repair the negative and harmful things we've done to the environment (Demetry 2022).

Responsive Interactivity

Similarly, responsive interactivity within the caring relationship was contested in the interviews. Interviewees admitted that giant sequoia trees do communicate chemically in response to environmental stimuli, like smoke which prompts physiological change in trees (Demetry 2022). These stimuli might be caused by humans, but there is no understanding on the part of the trees that a human caused this environmental change. If response is a form of communication, there is evidence that giant sequoia trees can communicate with one another through underground means. This is demonstrated in the swelling of stumps that are cut in groves, where sugars are transferred to the stump in the root system by other trees (Demetry 2022).

Interviewees discussed a variety of perspectives on communication and responsiveness. Some interviewees resisted the assertion that giant sequoia communicate with us humans (Demetry 2022). Giant sequoias, they say, do not really respond to individual humans. However, one interviewee discussed giant sequoia groves as a place of prayer where communication can and does occur directly, though carefully, between humans and giant sequoia trees (Coats 2022).

Communication is just one sign of responsive interactivity of many. One interviewee gave the example of symbiosis, where humans can use the bark of giant sequoias (usefully fire-resistant material) and humans provide controlled burns for the health of giant sequoias (Vera 2022). It is clear that giant sequoias can and do respond to environmental changes created by human actions.

Success Conditions of Care

Giant sequoias offer a clear case of the possibility of caring actions resulting in beneficial consequences for cared-fors. The history of the parks system attests to this: these lands were set aside to specifically protect these organisms from greed and unwise consumption. Caring actions can also fail to benefit the trees—case in point, failed fire-mitigation policies that result in hotter burning fires that kill off otherwise healthy trees and destroy seeds rather than helping them to germinate.

One common theme in my discussions with parks interviewees was the hope that visitors could connect with these trees and be motivated to help them by making changes in human behavior around climate change. This, along with the examples of symbiosis, reveals the belief that human actions can be beneficial for the outcomes of these trees. Human caring actions can provide benefits or harms for giant sequoias.

Conclusion: Paradigm Shifts in Caring for the Wild

Ideally under the preservationist or conservationist paradigms of the Parks Systems in which there is a clear dichotomy between domestic and wild, wild plants would need nothing from us. Our relationships could be casual and incidental, and there would be more than enough wild plants to meet the needs of not only humans, but more-than-human animals and other living organisms. Clearly, though, we live in a world where both the National Parks System and the National Forest Service are active managers of these wild landscapes. More importantly, the idealized removal of humans from these ecosystems is just one of many paradigms that are used to explain and make sense of both distinctions between the domestic and the wild, and humanity's role in either.

At best, perhaps the human-wild plant relationship could aim towards cultivation, as in Kat Anderson's ethnography *Tending the Wild*, when it can benefit both plants and humans (and even other animals).

> About halfway into the years of fieldwork, I began to ask native elders, "Why are many plans and animals disappearing?" Their answers, which always pinned the blame on the absence of human interaction with a plant or an animal, began to add up to a third major insight: not only do plants benefit from human use, but some may actually depend on humans using them.
>
> (Anderson 2013, xvi–xvii)

This approach is very compatible with care ethics.

From this approach, caring relationships are crucial for the thriving of humans and other organisms. The wild and domestic distinction does not cleanly account for cultivation, in which wild plants are tended to by humans. As Anderson explains:

> Human tending of certain California native plants had been so repetitive and long-term that the plants might very well have become adapted to moderate human disturbance. This idea had a very practical corollary: the conservation of endangered species and restoration of historic ecosystems might require the reintroduction of careful human stewardship rather than simple hands-off preservation.
>
> (Anderson 2013, xvi–xvii)

There is some interdependence between humans and wild plants, but not in the same way as colonial contexts where ecosystems are radically altered (e.g., industrial growing of domestic tomato plants) or plants radically changed (e.g., bonsai plants) to meet human desires, needs, or goals.

In the colonial context, our ancestors have created systems and habits that run on extraction. Most recently, that extraction centers on materials to create fossil fuels and technology (e.g., computer chips). Our human activities, intentionally or not, have a great influence on plants around the globe, both wild and domestic. Giant sequoia trees are no exception.

> The unprecedented number of giant sequoias lost to fire last year serves as a call to action. We know that climate change is increasing the length and severity of our fire seasons due to hotter temperatures and drought. To combat these emerging threats to our forests, we must come together across agencies. Actions that are good for protecting our forests are also good for protecting our communities.
>
> (Jordan 2021).

Our human actions intervene upon these beings through the changes our activities produce in the climate. In 2015, the Rough Fire affected Sequoia and King's Canyon National Park, "A total of 27 large sequoias were killed on park land in Grant Grove. In U.S Forest Service groves, at least 74 fire-killed large sequoias were documented" (National Park Service n.d.). One hundred twenty Giant Sequoias died in the 2019 Pier and Railroad fires, affecting the Sequoia National Forest and areas in Mariposa County. In 2020, things were much worse.

> The Castle Fire killed an estimated 7,500 to 10,600 large sequoias (those with trunk diameters of 4 feet or more, or ≥1.2 meters). This is 31 to 42 percent of large sequoias within the Castle Fire, or 10 to 14 percent of all large sequoias across the tree's natural range in the Sierra Nevada.
>
> (National Park Service n.d.)

And in 2021, the carnage continued.

> Taken together, the combined loss from the KNP Complex and Windy fires is estimated to be 2,261 to 3,637 large sequoias that have already been killed or will die in the next three to five years, or 3 to 5 percent of the sequoias in the entire Sierra Nevada population of large sequoias.
>
> (National Park Service n.d.)

To be clear, giant sequoia trees rely on fire for their successful procreation. Why, then, are so many dying? What has changed? The climate and our fire prevention practice (Dickman 2022).

In my conversations with caretakers, I asked if they had ever experienced the death of a giant sequoia under their care and what, if anything, it felt like. Interviewees recounted feelings of negativity and loss when a giant sequoia died. One described the dead tree as a "grey, ghostly skeleton" and felt "heartbreak" as they saw "one of the most depressing things to look at" (Benson 2022). Parks service employees interviewed are by and large scientists with advanced degrees and a great amount of experience in the field. So, they are used to seeing dead and dying plants; it is a biological reality in ecosystems that some living things die. What is different about this experience is the age of the organism, the change in speed of the decline, and the massive numbers dying. Deaths of giant sequoia trees, as mentioned above, have been speeding up, and contributing to a massive amount of loss.

Giant sequoia trees are not the only casualties of climate change. But as they are the focus of this case study, the emotional response towards their loss can be informative to our care ethic. Even with the view that these trees did not used to depend on us (Vera 2022), we are at a point now where at the very least require humans to "Get out of the way" and stop creating an uninhabitable and rapidly changing climate for these and other plants (Demetry 2022). The consequences of our activities driving climate change create more acute interdependency than was previously evident. This makes caring relationships and admitting plants into our ethics of care all the more pressing.

Note

1 Interviewee A (who asked to remain anonymous) used the terms "Native and non-Native" and "Indian." Language around identity of Native Americans/Indigenous folks/American Indians is fraught in the U.S.

References

Anderson, M. Kat. *Tending to the Wild: Native American Knowledge and the Management of California's Natural Resources.* University of California Press, 2013.

Benson, Teresa. Interview. 2022.

Bloom, Rochelle and Douglas Deur. "Ethnographic Data and Tribal Engagement at Yosemite National Park." *Land*, 9(9) (2020).

California Grants Portal. "Tribal Nature-Based Solutions with Tribal Fire Resilience." www.grants.ca.gov/grants/tribal-nature-based-solutions-with-tribal-wildfire-resilience/

Castillo, Ed. "Petition to Congress on Behalf of the Yosemite Indians." *The Journal of California Anthropology*, 5(2) (1978). https://escholarship.org/uc/item/72r473wg

Coats, Waylon. Interview. 2022.

Demetry, Athena. Interview. 2022.

Dickman, Garret. Interview. 2022.

Dickman, Garret. "Quote of the Day." *New York Times*. July 14, 2022. www.nytimes.com/2022/07/14/pageoneplus/quotation-of-the-day-firefighters-are-using-sprinklers-to-protect-a-treasured-giant-sequoia-in-yosemite.html

Fire Effects Information System (FEIS). "Sequoiadendron giganteum." USDA. www.fs.usda.gov/database/feis/plants/tree/seqgig/all.html.

Gibson, Prudence. *The Plant Thieves: Secrets of the Herbarium*. New South, 2023.

Hartesveldt, Richard J. *Giant Sequoia of the Sierra Nevada*. University Press of the Pacific, 2005.

Jacobs, Jamie. "The Great Law." Talk. Seneca-Iroquois National Museum, November 4, 2025.

Jordan, Clay. "Wildfires Kill Unprecedented Numbers of Large Sequoia Trees (U.S. National Park Service)." National Parks Service, July 2021. www.nps.gov/articles/000/wildfires-kill-unprecedented-numbers-of-large-sequoia-trees.htm (accessed May 31, 2022).

Katz, Eric. "The Rewilding Paradox: A World Free of Human Domination." *IaI News*, 2023.

Kimmerer, Robin Wall. *Braiding Sweetgrass: Indigenous Wisdom, Scientific Knowledge, and the Teachings of Plants.* Milkweed Editions, 2013.

Kohlruss, Carmen. "Conservation Group Seek to Block Logging Projects in Giant Sequoia National Monument." *Sierra: The Magazine of the Sierra Club.* February 23, 2024. www.sierraclub.org/sierra/conservation-groups-seek-block-logging-projects-giant-sequoia-national-monument

Leopold, Aldo. *A Sand County Almanac.* Oxford University Press, 1949.

McConville, Drew, Mariel Lutz, and Jenny Rowland-Shea. "The Trump Administration's Expansive Push to Sell Out Public Lands to Highest Bidder." *Center for American Progress*, September 22, 2025. www.americanprogress.org/article/the-trump-administrations-expansive-push-to-sell-out-public-lands-to-the-highest-bidder/

Molina, Felipe S. and David Delgado Shorter. "Episode 47 Felipe S. Molina and David Delgado Shorter." Networking with Plants in the Anthropocene Podcast. November 1, 2025. https://soundcloud.com/networking-with-plants/episode-47-felipe-s-molina-and-david-delgado-shorter

Muir, John. *Our National Parks.* Houghton Mifflin, 1901.

National Parks Service. "Wildfires Kill Unprecedented Numbers of Large Sequoia Trees." n.d. www.nps.gov/articles/000/wildfires-kill-unprecedented-numbers-of-large-sequoia-trees.htm

NDN Collective. "Landback." https://ndncollective.org/landback/

Save the Redwoods League. "California's Giant Trees." n.d.

Spence, Mark. "*Dispossessing the Wilderness: Yosemite Indians and the National Park Ideal, 1864–1930.*" *Pacific Historical Review*, 65(1) (1996), 27–59.

Taylor, Paul W. *Respect for Nature: A Theory of Environmental Ethics*, 25th anniversary edn. Princeton University Press, 2011.

Tweed, William C. *King Sequoia: The Tree That Inspired a Nation, Created Our National Park System, and Changed the Way We Think about Nature.* Sierra College Press, 2016.

Tweed, William C. Interview. 2021.

University of Washington School of Law. "Sovereign-to-Sovereign Cooperative Agreements." 2024. https://lib.law.uw.edu/c.php?g=1329421&p=10063577

USDA. "Sequoia National Forest – Home." www.fs.usda.gov/detail/sequoia/home/?cid=stelprdb5394941 (accessedAugust 23, 2022).

U.S. National Parks Service. "Exploring Giant Sequoia Groves - Sequoia & Kings Canyon National Parks." www.nps.gov/seki/planyourvisit/sequoiagroves.htm (accessed August 23, 2022a).

U.S. National Parks Service. "History & Culture – Sequoia & Kings Canyon National Parks." www.nps.gov/seki/learn/historyculture/index.htm (accessed June 3, 2022b).

Vera, Kerri. Interview. 2022.

Yosemite National Park. "Their Lifeways." www.nps.gov/yose/learn/historyculture/their-lifeways.htm

Westover, Robert Hudson. "Conservations versus Preservation?" USDA Blog. 2016. www.usda.gov/about-usda/news/blog/conservation-versus-preservation

Willard, Dwight. "*The Natural Giant Sequoia (Sequoiadendron Giganteum) Groves of the Sierra Nevada, California-An Updated Annotated List.*" n.d.

Conclusion

Caring for Plants

We talk about plants being cared for in our vernacular. Is this mere custom or does our care of plants have moral weight? Care ethics seems especially well suited to engage with the inclusion of plants and other environmental entities in part because of its rich history navigating a relationship between descriptive areas of study (e.g., psychology and anthropology) and prescriptive ones (e.g., ethics). The work of early care ethicists was prompted by scientific observations that suggested a difference in behavior and thinking amongst genders (Kittay 2019, 168–9). Engagement with empirical research is sometimes necessary in feminist philosophical spaces to motivate skeptics to acknowledge that the phenomena feminists identify are real and worthy of attention. However, extensive use of empirical work within philosophical spaces can open one up to concerns about whether the research and conclusions are really philosophical in nature, and not merely part of a more descriptive endeavor. Care can be such an area of difficulty. It is clear that descriptively humans care about and for many things. Care work is required for our physical existence, as all of us reading this were at one point infants unable to navigate our environments and survive without intensive support from other humans. But why should this practice, care, be elevated to the stuff of our moral orientation towards the world? I think care ethics faces a "partners in crime" scenario, where upon close examination of any other normative ethical there becomes a point where we might think, "but really, why should *this* be the basis of our moral obligations?" For all of its issues, care ethics seems at least as plausible as other approaches (e.g., utilitarianism, deontology, virtue ethics, etc.)

In the case of caring for plants, what actually changes if we admit a moral obligation to care for plants? As mentioned earlier, I do not think this means that we as individuals have an obligation to care for all plants at all times because we are not in caring relationships with all plants at all times and we are not in a position to care for all plants at all times. Yet, all plants are potential cared-fors because they have inherent moral value and the capabilities and capacities relevant to an ethically caring relationship. So, they are

DOI: 10.4324/9781003603665-9

available for care when the context calls for it. We are obligated to begin acknowledging the plants in our lives as potential cared-fors. The houseplants, vegetal food, and plant ecosystem members are alive in our moral worlds in different way than they were before.

I would argue that anthropogenic climate change is such a context that illuminates the obligation of human carers to care for plant cared-fors. Not only is the activity of many humans (historical and contemporary) undermining the flourishing of other humans and other animals, but also plant cared-fors. While it may not appear like much when looking at our other ethical obligations to support the flourishing of so many entities, admitting plants as cared-fors re-orients us to the members of our communities that matter.

In this project, I demonstrated the initial barriers to including plants within our ethics of care. Primarily, the barriers relate to the human-centered foundation of contemporary care ethics and who is admitted into the caring relationship as a cared-for. Admitting other animals is controversial enough and admitting plants as cared-fors presented major challenges.

In considering options for the inclusion of plants in care ethics given the structural elements of ethical care that favor beings like humans, I offered five possible paths forward. Plants, under any of those accounts, have inherent moral value. The capabilities they have as living organisms also provide a good basis for inclusion as cared-fors.

Three concerns that I especially wanted to address including plants in care ethics as cared-fors is

1 the ability of the carer to have valuing-attitudes for the plant cared-for,
2 the existence of a caring relationship that can include caring affect, dependency, and responsive interactivity, and
3 success conditions that demonstrate the care can be taken up by the cared-for and contribute to their well-being (i.e., be beneficial to them).

I argued that both human carers and plant cared-fors are able to function in an ethically caring relationship in light of these concerns.

The case studies exist as practical application and a precursor (or initial step into) field philosophy (Brister and Frodeman 2020). Developing the case studies required interviews, because it was important to me to actually talk with plant practitioners about their attitudes, feelings, and practices. This approach better ensured that I was understanding what their practice is actually like according to them, rather than simply reading their work and possibly misinterpreting their views. On a personal note, while writing is wonderful, I also learned how important and useful it is to provide a space for practitioners to discuss things they think about as they work with plants. Their experiences are valuable, and taking the time to connect with them and learn from them was invaluable. It made me realize the importance of philosophy outside of the classroom, research office, library, academic conference, and armchair. While these case studies may appear overly descriptive, I hope that my intent to follow in the lineage of what I take is one important characteristic of environmental philosophy—to consider and apply one's theoretical considerations and arguments to the real world. Experiences with our environment—plants, other animals, landscapes, ecosystems—have prompted revisions of who or what we find valuable, reevaluations of the consequences of our actions, and reconsideration of our human identities. While this application is not nearly so important, it is done in the spirit of the cooperation and dance between theory and practice that appears throughout environmental philosophy.

This project merely scratches the surface on the issue of plants and care ethics. But I hope it has been a useful first step in considering the issues that need to be address to include plants as cared-fors within care ethics.

This Project and the Direction of the Plant Humanities and Critical Plant Studies

In the years since this book has been in development, the terrain of the plant humanities and critical plants studies has changed greatly. I began researching this topic around 2017. Since then, a wide breadth of research has been produced. I think that skepticism remains amongst folks whose worldviews (scientific, religious, cultural, or otherwise) do not admit for plants having complex capacities and capabilities. But the options available to the person interested in integrating plants into existing frameworks through extension or revision has grown in the literature.

Now there is a more radical push to accept an even stronger and more profound relational ethic beyond the vision of care ethics I provide here. I mentioned in the beginning of this book that the question of intrinsic value goes in and out of fashion within environmental ethics. One might question whether my account (or any care ethics account) could make sense without such moral value, which I seem to center in this analysis. In some way, relevant capabilities or capacities are the very thing that might qualify someone or something as having intrinsic moral value. As mentioned in Chapter 2, the capability of autonomy is keenly tied to personhood, a whole range of capacities from feeling emotion to having a sense of the future qualifies one as a subject-of-a-life, being alive as a capability of a living organism, and the capacity of flourishing. If those capacities or capabilities are characteristics or indicators of intrinsic moral value, why take the extra step to assert that value when the capacity or capabilities make sense of what is required for a caring relationship? Given this, one might question my insistence on both intrinsic moral value and care relevant capacities or capabilities.

Another variation of this project might focus on the components of the caring relationship itself, regardless of identity markers within the carer and cared-for. I think that such an approach is possible and it is a project I am looking forward to pursuing in the future. What if care ethics were more focused on the relationship itself rather than the players in the relationship?

What if moral value were rooted in the relationship itself, and the identities and capacities of the individuals or groups engaged in the relationship were less crucial for establishing the aptness of the theory? That would be a somewhat radically different project, and one that some in the plant humanities are moving towards. Rather than try to fit plants into our human-centered and human-filtered understanding of the world and morality, perhaps we should look to the stark differences between humans and plants, to develop a morality that is more conducive to including more than animals, if not anchoring such a view in the radically other. I think this approach is worthwhile, even though it is not an approach I take here. Coming from the traditions that I do, initially my goal was to build a bridge from human centered ethics to the more-than-human world. I chose to do this through a clearly extensionist approach. Extensionist approaches are controversial, however. In some cases, this concern about extension is well-founded. As discussed extensively in Chapter 2, conceptual tools like personhood emerge from human communities thinking about how to value other humans and innately human activities. If you extend them too far, they make more-than-human beings look less like themselves, and more like humans.

One might be concerned that extending concepts developed by and for humans (e.g., personhood, care, ethics) might be a case of pernicious anthropomorphism. While mentioned briefly in Chapters 2 and 3, anthropomorphism is lurking seriously in the shadows of this project. Anthropomorphism has a checkered existence in philosophy. Often vilified as a tool for either bending the more-than-human to fit the mold of the human, or unjustifiably projecting the human image onto the more-than-human, anthropomorphism is complex. In some cases, it could be praised for assuming similarity that grants moral consideration to those we may not know for sure are conscious or sentient. In the case of animal ethics, at least the initial move of granting social more-than-human animals the bonds of friendship could be considered very anthropomorphic, even though it might be the best way to accurately describe those social bonds. While scientific paradigms have (hopefully) shifted to the point where it is not on-its-face ridiculous to claim that cows have best friends (Gutmann, Špinka, and Winckler 2015) or that elephants mourn the dead (Monsó 2024), the case of assigning such descriptors to plants is even more controversial. Plants exist very differently than humans. Rarely will you find someone who would deny the stark differences between humans and plants. Yet, the initial projecting of experiences that we, as humans, are used to, onto plants may not be the worst thing imaginable. I think it is what happens after that projection. What initiates the projection? What follows it? Does the human person involved seek out more understanding, and test the hypothesis of a particular plant having a particular capability? We humans clearly have similarities with plants (we have evolved together on this planet, after all). We also clearly have differences. In this space of navigating the similarities and differences, anthropomorphism should not be immediately rejected. Rather, one can proceed with caution, self awareness, and a critical eye towards evaluating the justification of anthropomorphic claims or observations.

In future research, I would like to extend care ethics into different contexts involving different plants. For example, how does ethical care work when considering wild plants that humans are barely aware of in their environments? And what about weeds? Another area largely avoided in this work is the popular topic of native and non-native plants. Discrimination amongst types of plants and individual plants as they relate to our caring relationships with them is a reasonable future investigation in considering the application of care ethics.

Also, while in this text I attempted to avoid particular belief sets about plants and plant capacities and capabilities underlying individual practitioners' work with plants, I think investigating the relationship of these beliefs to practitioner actions is an area of interest for future work. Does how we understand plant flourishing relate to the success conditions of the cared-for plant? This is a project that would engage deeply with empirical research. By providing rich QCS analysis to support philosophical investigations of how these belief states relate to success conditions for the cared-for.

Another area that needs to be expanded in future work is a deep dive into what a meaningful or worthwhile life for a plant is. Is plant well-being the same thing as a plant's welfare? Does a plant have a welfare? Is plant flourishing simply plant well-being? Or is plant flourishing something altogether different than mere health? Does having relationships with other living and non-living components of its environment matter to the plant? If it does, why? For its biological fitness or for some other value being in these relationships has for the plant? These are things to consider taking this project further, especially as the field of critical plant studies is working to connect plant science and plant humanities in new and interesting ways. There is so much more to explore.

A General Ending Note: Climate Change and Caring Relationships

Part of what prompted this research was personal concern about climate change. While this existential fear of mass destruction and the disappearance of human beings may not have been visible on the surface of the project, it is part of what underlies a motivation for finding a way to include other-than-humans in our moral frameworks. Climate change is such a major moral issue even without considering other beings, or only considering them instrumentally related to own species. But, when we imagine solutions and cope with environmental shifts and experience loss and grief in light of current and impending climate related devastation and suffering, it seems essential to include more-than-humans in our considerations. One of the feelings that has most moved me in this research and life in the wake of this project is grief. The grief experienced by people who have worked with plants is revealing to me. It goes beyond the loss of a particular being (e.g., your favorite bonsai plant, the end-of-season death of your tomato plant, or a giant sequoia tree you admired). I believe it is the loss of the relationship. Of course, part of that relationship is the existence of the other party that you're in relationship with. But the loss of that relationship includes more: the care protocols that measure and moderate your life. The time set aside to spend with that plant. The gifts that the plant provides, physical in the exchange of CO_2 for oxygen, emotional in the prompting of feelings in your time spent with it, and possibly spiritual, depending on your belief sets. These all take place in relation. And while many might view themselves as Simon and Garfunkel's proverbial island, human existence is at its core relational; whether we acknowledge it or not. Our grief is indicative of the loss of these valuable relationships. Many of the deaths and cessations of relationships are due, at least in part, to human activity. While this isn't a project on responsibility, I think the acknowledgment of these relationships and the orientation of care requires us to act. As philosopher Yogi Hendlin notes, feeling is key to existing in this world as human beings (Hendlin 2025). We need to be present in this grief, and while experiencing it, realize what it says about what we value. Importantly, care ethics has a story to tell about the importance of emotions in our ethical lives. These feelings can point us to what is at stake if we do not act now, in whatever capacity we can, to restore caring relations with plants in our world. This restoration requires admitting our vulnerable state (something that care ethics does particularly well). Human beings are essentially dependent on others for our existence, survival, and success. From that position of extreme interdependency, we can accept our need for receiving care and our duty to provide care. I hope we do. I hope we care.

References

Brister, Evelyn and Robert Frodeman. *A Guide to Field Philosophy: Case Studies and Practical Strategies.* Routledge, 2020.

Gutmann, Anke Kristina, Marek Špinka, and Christoph Winckler. "Long-term familiarity creates preferred social partners in dairy cows." *Applied Animal Behaviour Science*, 169 (2015), 1–8.

Hendlin, Yogi. "Biosemiotics: A New Way to Understand Non-Human Consciousness." *Essentia Foundation.* Video Interview. www.youtube.com/watch?v=6hHRtu8Y2sM. 2025.

Kittay, Eva Feder. *Learning from My Daughter: Valuing Disabled Minds and Caring That Matters.* Oxford University Press, 2019.

Monsó, Susana. *Playing Possum: How Animals Understand Death.* Princeton University Press, 2024.

Index

The letter n following a page number indicates a reference in the notes.

For Product Safety Concerns and Information please contact our EU
representative GPSR@taylorandfrancis.com
Taylor & Francis Verlag GmbH, Kaufingerstraße 24, 80331 München, Germany

www.ingramcontent.com/pod-product-compliance
Ingram Content Group UK Ltd.
Pitfield, Milton Keynes, MK11 3LW, UK
UKHW021307260926
475702UK00048B/798

* 9 7 8 1 0 3 2 9 9 3 4 7 8 *